Platforms and the Planet

DIGITAL MATERIALITIES AND SUSTAINABLE FUTURES

Series Editors: Adi Kuntsman, Manchester Metropolitan University, UK.
Erinma Ochu, The University of the West of England, Bristol, UK.
Liu Xin, Karlstad University, Sweden.

The materiality of digital technologies can inflict substantial environmental damages: the ever-growing extraction of resources needed to produce digital devices; the toxicity of e-waste; and the rapidly increasing energy demands of data centres, needed to sustain every click, website, database or 'smart' network. And yet, the geography of digital environmental harms is unevenly distributed and is often aligned with geographies of colonialism, land dispossession, extraction of natural and human resources, and the violence of industrialisation and capitalism. As digital economies grow at an unprecedented speed, the links between global environmental degradation and digital industries become more apparent, as does their intersection with global environmental injustice. However, when these questions are addressed, they are typically framed as problems of cost and efficiency where the environment is simply seen to provide material condition for the operation of digital economy.

Digital Materialities and Sustainable Futures develops an interdisciplinary field that shifts the conversation about digital environmental sustainability from profit and efficiency to geographies of in/justice and the colonially and decolonisation of digital scholarship and practice.

EDITORIAL ADVISORY BOARD

Platforms and the Planet: Big Tech, Digital Platforms and Environmental Responsibility

EDITED BY

SALLA-MAARIA LAAKSONEN
University of Helsinki, Finland

MERVI PANTTI
University of Helsinki, Finland

AND

OLGA DOVBYSH
University of Helsinki, Finland

emerald
PUBLISHING

United Kingdom – North America – Japan – India – Malaysia – China

Emerald Publishing Limited
Emerald Publishing, Floor 5, Northspring, 21-23 Wellington Street, Leeds LS1 4DL

First edition 2026

British Library Cataloguing in Publication Data
A catalogue record for this book is available from the British Library

ISBN: 978-1-83662-173-7 (Print)
ISBN: 978-1-83662-172-0 (Online)
ISBN: 978-1-83662-174-4 (Epub)

Printed and bound by CPI Group (UK) Ltd, Croydon, CR0 4YY

INVESTOR IN PEOPLE

Contents

Section 2: Infrastructures and Materiality

Section 3: Discourses of Responsibility

Section 4: Closing Words

About the Editors

Dr. Salla-Maaria Laaksonen is a Senior Researcher at the Centre for Consumer Society Research and an Adjunct Professor of Media and Communication Studies at the University of Helsinki. Her research focuses on communication and organising in the platform society, including social evaluation of organisations in the hybrid media system, the organisation of online social movements and the use of data and algorithms in organisations. She is also an expert in digital research methods with solid experience in computational social science research projects and research use of social media data.

Mervi Pantti is a Professor in Media and Communication Studies at the University of Helsinki, Finland. Her research concerns media responsibility, relationship between media and emotion, conflict and crisis reporting and epistemic rights of marginalised groups. She is working on a research project on media and democratic epistemic capacities.

Dr. Olga Dovbysh is a postdoctoral researcher at the Centre for Consumer Society Research, University of Helsinki. Her research focuses on media and internet control in restricted media regimes, journalistic cultures and environmental and climate communication. She is a co-founder of the academic initiative 'Online Talks on Eurasian Media', aimed at bringing together researchers and journalists to discuss current and emerging problems in the field of media and communication across the vast territory of Eurasia. Dovbysh's recent publications have appeared in journals such as *Journalism*, *Digital Journalism*, *Media & Communication and Social Media + Society.*

About the Contributors

Flora Mary Bartlett is a visual and environmental anthropologist and Assistant Professor at The Department of Thematic Studies – Technology and Social Change, Linköping University. In her research, she uses creative visual methodologies to focus on landscape relations and environmental challenges in rural Sweden. She often works with exhibitions as part of her research practice, collaborating with different museums and co-curating the 2025 exhibition *Where is the Cloud?* together with Julia Velkova.

Lauren E. Bridges is an Assistant Professor of Media Studies at the University of Virginia and a faculty affiliate at the Berkman Klein Centre for Internet & Society at Harvard University, U.S.A. She researches the sociotechnical, political economic and environmental politics of digital infrastructures. Dr. Bridges has published in peer-reviewed journals, public news outlets such as *The Guardian*, and she is regularly interviewed in media regarding the social and environmental impacts of big data infrastructures. She is currently writing a book on the local land use politics of digital industrial expansion in Southern California and Northern Virginia.

Tessa J. Brown is a PhD student in Cultural Studies at Queen's University (Canada) and co-editor of Heliotrope (heliotropejournal.net).

Simon Cottle (PhD) is a Professor Emeritus, Media and Communication, in the School of Journalism, Media and Culture (JOMEC) at Cardiff University where he was formerly the Deputy Head and Head of School. Simon is the author of 14 books and over 150 chapters and articles, many on global crisis reporting, and is the Series Editor for the Global Crises and the Media Series for Peter Lang Publishing. His latest book is *Communicating a World-In-Crisis* (2025). He is currently writing and lecturing on journalism and our world-in-crisis, performs as a guitarist and writes songs on ecology and the state of the planet.

Meri Frig is a postdoctoral researcher at Aalto University, School of Business, specialising in corporate social responsibility (CSR) communication and business-society relations. Her interdisciplinary research focuses on the roles of cultural and information intermediaries in addressing the sustainability crisis. Her work has been published in journals such as the *Journal of Business Ethics*, *Business & Society* and *Media, Culture & Society*.

Steph Hill is a lecturer at the University of Leicester in the United Kingdom. She is the chair of the Spinning Climate Change Working Group at the Climate Social Science Network and the Co-chair of the Emerging Scholars Network at the Climate Social Science Network. Her work examines governance processes on platforms and in environmental policy with a focus on activism that targets businesses. Her current research project investigates the negotiations of a global plastics treaty at the United Nations Environment Assembly.

Mél Hogan is an Associate Professor in Film & Media at Queen's University (Canada). Her research focuses on data infrastructure, as understood from within the contexts of planetary catastrophes and collective anxieties about the future. Dr. Hogan is the PI of the Genomic Media project, host of The Data Fix podcast (thedatafix.net) and editor of Heliotrope (heliotropejournal.net).

Elis Karell is a doctoral student in Media and Communication Studies at the University of Helsinki, Finland. His research interest is in climate change communication, particularly environmental sustainability discourses and climate technologies.

Tero Karppi is an Associate Professor of Critical Computation and Digital Media at the University of Toronto, Canada, and a visiting researcher at the School of Engineering Sciences, Department of Social Sciences, LUT University, Finland. Karppi is the author of *Disconnect: Facebook's Affective Bonds* (Minnesota, 2018), one of the authors of *Undoing Networks* (Minnesota/Meson, 2021), and one of the editors of *Reckoning with Social Media* (Rowman & Littlefield, 2021).

Martina Skrubbeltrang Mahnke serves as an Associate Professor at Roskilde University, Denmark. Her research focuses on how developers, content creators and users (co-)create digital platforms and what this entails for contemporary societies and the planet. Positioned at the intersection of digital communication studies, datafication and platform studies and sustainability studies, her work aims to unpack the social and environmental implications of digital platforms through the development of new methodologies and conceptual frameworks. Prior to this, she has mainly written about the multifaceted sociotechnical relationship(s) on social media and eHealth platforms, and she has studied the utilisation of generative AI in work-related communication contexts.

Vusumuzi Maphosa is the Information and Communication Technology Department Director at the National University of Science and Technology, Zimbabwe. He has published book chapters, peer-reviewed journal articles and conference presentation posters. Vusumuzi is co-PI a multinational research grant from the Internet Society Foundation (ISOC) that focuses on discard challenges affecting the Global South. Vusumuzi's research covers transformative technological concepts, including ICT4D, educational technology, Industry 4.0, data mining, artificial intelligence, semantic webs and e-governance.

Markus Ojala is a University Lecturer in media and communication studies at the University of Helsinki, Finland. His research has explored several topics in political communication, including the organisation of online publics around immigration, the public's trust in the news media and the media coverage of international crises and conflicts.

Thomas Olsson is a Professor in Human-Centred Design at Tampere University, Finland. He specialises in sociotechnical systems and computer-supported cooperative work, seeking to understand the diverse interconnections between emergent information technology and complex social processes. Recently, Olsson has investigated sociotechnical applications from the perspectives of environmental and social sustainability, long-term and indirect repercussions and design responsibility. This chapter builds on his recent curriculum development work on Sustainable Design, aiming to steer the praxis of designing digital services towards a culture that prioritises sustainability.

Gopika Premsankar is an Assistant Professor in the Department of Information and Communications Engineering at Aalto University, Finland. She researches networked systems, with a primary focus on enhancing sustainability and performance. Situated at the intersection of wireless communications, cloud and edge computing and network infrastructure, her research employs tools from optimisation, applied machine learning and statistics to achieve these goals.

Olli Pyyhtinen is a Professor of Sociology and the founder of Relational Studies Hub (RS Hub) at Tampere University, Finland. His research intersects social theory, philosophically inclined fieldwork, science and technology studies, economic sociology and the study of art. Pyyhtinen is the author of, for example, *More-than-Human Sociology* (Palgrave, 2015), *The Simmelian Legacy* (Palgrave, 2018), *The Gift and its Paradoxes* (Routledge, 2014) and *Simmel and 'the Social'* (Palgrave, 2010). Currently, he is leading two projects on waste and the circular economy, WasteMatters (ERC Consolidator Grant, 2022–2027) and DECAY (Research Council of Finland, 2022–2026).

Julia Velkova is a Professor of Media and Culture at Linköping University and a ProFutura Scientia Fellow at the Swedish Collegium for Advanced Studies. Her work explores the entanglements of digital infrastructures with energy politics and the temporalities of large scale media infrastructures. She is co-editor of the book *Media Backends: Digital Infrastructures and Sociotechnical Relations* (with Lisa Parks and Sander De Ridder, 2023) and of *Beyond Academic Publics: Conversations about Scholarly Collaborations with Cultural Institutions* (with Anne Kaun, 2024).

Deb Verhoeven is Canada 150 Research Chair in Gender and Cultural Informatics at the University of Alberta (Canada). She is Co-PI of the Genomic Media project.

Minna Vigren is an Assistant Professor of Global Communications and Climate Change at LUT University. Her interdisciplinary research bridges critical internet studies, science and technology studies, human–computer interaction, radical pedagogy and futures studies. Her work focuses on themes such as digital sustainability, alternative futures, imagination, power and resistance. She currently leads the project *Imagining Sustainable Digital Futures: Outlines for Eco-Digital Citizenship* (Research Council of Finland, 2022–2025).

Acknowledgements

Platforms and the Planet: Big Tech, Digital Platforms and Environmental Responsibility (Emerald Publishing, 2025), edited by Salla-Maaria Laaksonen, Mervi Pantti and Olga Dovbysh.

The initial idea for this book was born in watercooler discussions related to a research project focused on the accountability of digital media platforms. In the middle of growing awareness of the climate crisis as a global, escalating disaster (Ripple et al., 2024), we wanted to turn our gaze to the complex issue of environmental sustainability of digital technology and most notably, the platform companies. Luckily, we did not have to start from scratch, but we were able to reach out to many scholars who had already been pioneering in the field.

This book brings together a marvellous group of scholars across the globe. The original ideas for the chapters were based on ideas presented by the authors themselves, and they are based on ongoing empirical work and critical thinking. We would like to express our gratitude to all the authors who patiently applied and responded to our suggestions and comments. We also owe to the external reviewers who commented on some of the chapters, as well as to the editorial team at Emerald. As editors, we learned a lot during the process, and we sincerely hope our readers will have the same experience!

This work was financially supported by the Research Council of Finland [grant number 332751] and the Strategic Council (SRC) established within the Research Council of Finland [grant number 372226], two project grants by Kone Foundation: FLOWISION and Greener Bits and a Catalyst Grant funding from Helsingin Institute for Social Sciences and Humanities.

References

Ripple, W. J., Wolf, C., Gregg, J. W., Rockström, J., Mann, M. E., Oreskes, N., Lenton, T. M., Rahmstorf, S., Newsome, T. M., Xu, C., Svenning, J.-C., Pereira, C. C., Law, B. E., & Crowther, T. W. (2024). The 2024 state of the climate report: Perilous times on planet Earth. *BioScience*, *74*(12), 812–824.

Introduction: Conceptualising and Theorising the Environmental Responsibility of Digital Platforms

Salla-Maaria Laaksonen, Mervi Pantti and Olga Dovbysh

University of Helsinki, Finland

Abstract

Digital technology is often considered a green and sustainable alternative and a solution to the climate crisis. However, emerging academic and public discussions highlight that digital platforms and the devices on which we use them have severe environmental consequences through, for instance, their escalating energy consumption, material resource use and the production of hazardous waste. This book seeks to develop a critical understanding of the environmental responsibility and accountability of digital platforms. It focuses on the environmental responsibility of digital media platforms and their role in the environmental sustainability transition as a discursive, material and ethical question. Overall, the book bridges critical platform studies with environmental studies and environmental communication studies, providing a multifaceted view of the dynamic relations between digital platforms and the environment. Empirically, the book also expands the perspective beyond Western platform companies by including evidence from the Global South, China and Russia for a more nuanced understanding of the platforms' power in relation to the planet.

Keywords: Digital media platforms; environmental sustainability; platform power; social accountability; digital materiality; responsibility discourses

Introduction

Digital technology is often considered an enabler and driver for solving major global challenges, such as climate change or the sustainability transition.

Platforms and the Planet, 1–18

doi:10.1108/978-1-83662-172-020261001

Likewise, digital media platforms, social media services offered by the so-called Big Tech companies, have been considered to offer a less wasteful way to consume media and participate in society through public communication. However, emerging academic and public discussions highlight that digital platforms, the devices on which we use them and the growing data centre industry are contributing to the climate crisis through, for instance, their escalating energy and water consumption and the production of hazardous waste (e.g., Arcieri, 2022; Brevini & Doctor, 2023; Edwards et al., 2024; Freitag et al., 2021; Hogan, 2018; Istrate et al., 2024; Lucivero, 2020; Parikka, 2015). The share of greenhouse gas (GHG) emissions of the entire information and communication technology (ICT) sector in terms of global emissions has exceeded that of the widely discussed air traffic sector, and the share is estimated to multiply in the coming years (World Bank & ITU, 2024). This is due to the downstream emissions caused by the growing use of digital services as well as increasingly computation-heavy technologies such as artificial intelligence (AI; see Crawford, 2021). Similarly, platform companies' emissions continue to rise despite their increasing investments in green technologies (Arcieri, 2022). This phenomenon is known as the Jevons paradox: as technological advancement increases efficiency, resource consumption rises (Alcott, 2005).

Despite the increasing awareness, the environmental responsibility of digital platforms is a topic that has been only rarely explored by scholars (Kannengießer & McCurdy, 2021; Kuntsman & Rattle, 2019; Kuntsman, 2020). The environmental footprint of digital technologies and platforms is difficult to observe, trace and measure (Laaksonen & Premsankar, *this volume*). The environmental impacts of digitalisation are often distant from a consumer's perspective: the digital services can be produced in data centres on the other side of the globe, and the energy required for data transfer is not made tangible to the consumer charging their smartphone. The e-waste resulting from outdated consumer devices and servers is exported to third-world countries, where hazardous materials cause environmental and health damage to local communities. In economic and environmental studies, the concept of environmental externality has been classically used to denote the side effects that impact third parties outside the primary economic activity (Buchanan & Stubblebine, 1962). In the platform economy, the environmental consequences are formed in a global ecosystem where harms are externalised from the Global North, who are the main consumers of digital services (e.g., Andreucci et al., 2023; Brodie, 2023; Parikka, 2015) and who are, more generally, responsible for the majority of GHG. While the impacts of climate change manifest globally, developing countries in the Global South face disproportionate risks of climate change–related natural disasters and threats to public health and lives. The inaction of developed nations and their disregard for the Global South reinforce existing and create new inequalities and colonial-era power dynamics (Mejias & Couldry, 2024; Pantti & Karell, *this volume*).

The platforms themselves mostly remain silent about the environmental harm and injustice related to their actions. Instead, the amalgamation of green and digital evident in the societal discourses on green and digital transition is also prominent in the ways in which technology companies themselves construct their social and environmental responsibilities: their sustainability reports, news releases and blog posts talk about their efforts to build a greener world (Frig et al., *this volume*). Overall, in the corporate and political sustainability discourse, the development of digital technologies is often approached from a solution-centric perspective, assuming that there are unproblematic technological solutions to societal problems (Kuntsman & Rattle, 2019; Morozov, 2014). Consequently, digital technologies are often presented as a means to save energy and reduce carbon emissions, even if current evidence suggests that the use of digital technology is growing faster than its energy efficiency (Freitag et al., 2021).

This edited volume focuses on the environmental responsibility of digital media platforms and their role in the environmental sustainability transition as a discursive, material and ethical question. Taking a critical and cross-disciplinary perspective, the book centres on platforms because of their global infrastructural power and their growing role in climate politics: not only as a communication infrastructure enabling our climate-related discussions and our digital economy but also as corporations with a political and moral responsibility for the development of our society. Therefore, we argue that it is essential to address digital platforms, and the Big Tech companies offering them, as key infrastructural and political actors in environmental sustainability, as well as to study their environmental impact in the context of rapidly increasing digital media consumption.

To unravel the complex relations that constitute the environmental responsibility of digital platforms, the chapters in this book seek to engage with this topic from three broad perspectives: first, by laying out the premises and causes of the environmental (un)sustainability of digital platforms; second, by unpacking the digital materiality of platforms as an infrastructure; and third, by investigating the discourses of responsibility around platforms and environmental sustainability. Therefore, while our focus is on the somewhat abstract notion of 'platform companies', the perspectives covered in the book also include other actors in the platform assemblage: users, activists, technology designers, regulators, as well as the broader society increasingly intertwined with the platforms. Yet, in the picture we paint together with all the chapter authors, the main targets of accountability demands are the platform companies.

In this introduction chapter, we first lay out the premises for platform power when it comes to environmental sustainability: their infrastructural power position in our society and the context of platform capitalism. Next, we discuss the material aspects of the digital platform society, and how they feed into the importance of studying platform accountability both as an ethical and discursive question. Finally, we present the book content by summarising each section and chapters that belong to it: (1) unpacking environmental sustainability, (2) infrastructures and materiality and (3) discourses of accountability.

Platforms Are Powerful – And Capitalistic

The role of transnational platform companies as providers of digital media infrastructure has been widely recognised (e.g., Plantin et al., 2018; van Dijck, 2021). These companies are now argued to hold major infrastructural power positions in society through their role in not only mediating public communication and occupying people's everyday lives (Helberger et al., 2018; Plantin et al., 2018) but also in deeper transformations of social practices and relations of individuals and organisations (e.g., van Dijck et al., 2018). Platform companies, as controllers of digital infrastructures, can influence the behaviours of other actors and markets by setting standards and rules and moulding the institutional environment (Khanal et al., 2024; Laaksonen et al., 2024; Nieborg et al., 2024; Srnicek, 2017; van Dijck et al., 2018). In scholarship, this development and form of power is typically called platformisation (e.g., van Dijck et al., 2018).

Further, in their capacity as informational intermediaries, digital platforms exercise a unique form of opinion power, which enables them to command online attention, set the agenda and use data and algorithmic tools to steer public discussion (Helberger, 2020; van Dijck & Poell, 2015). Instead of a public space shared by all, social media generates personal 'public sphericules' that cater for the interests of the particular user (e.g., Willson, 2014), supported by platform algorithms programmed to support the business logic of the platforms. Also, the public discussions on environmental sustainability and related activism are subject to these algorithmic logics, which highlights platforms' power in shaping our shared understandings of the climate crisis.

Apart from playing a major role in the distribution and circulation of media content, some platform companies also have a more hidden role as providers of server infrastructure and computing services (notably Amazon, Google, and Microsoft). By storing the data and running the software and calculations of other actors in the digital economy, these players play a powerful role as the backbone of the entire platform society. Digital technology is the pervasive infrastructure of the modern economy (Srnicek, 2017). Once adopted and integrated, leaving these infrastructures is made difficult, which further consolidates their power (Bartlett & Velkova, *this volume*; Brevini & Swiatek, 2020; Lehdonvirta, 2022). Arguably, such power is even more difficult to curb and regulate, and nearly impossible to circumvent as an individual user. It has been suggested that infrastructural cloud arrangements, specifically, drive market domination and rapid expansion in the platform economy (Narayan, 2022). The capital accumulation of platforms is built around data extraction and control, as well as network effects based on a growing user base, which further supports the centralisation of power and the development of a monopolistic market (e.g., Narayan, 2022; Srnicek, 2017).

Platforms Are Exploitative

Platforms feed on data. In platform capitalism, quality data are essential for the business models of the platforms (Srnicek, 2017; Van Dijck et al., 2018). Just as capitalism was made possible by the extraction of cheap natural resources (e.g., Moore, 2015), platform capitalism is made possible through the extraction and appropriation of seemingly free and intangible data from human users (Narayan, 2022). These datasets are born through data relations, where human relations are appropriated and datafied for commodification (Alaimo & Kallinikos, 2024; Couldry & Mejias, 2018; Mejias & Couldry, 2024). Couldry and Mejias (2018) call this novel form of colonialism *data colonialism* and argue that by generating such data relations, social media platforms reinforce existing colonial power structures. A central tenet of this development is the normalisation of exploitation of human beings through data, 'new type of appropriation that works at every point in space where people or things are attached to today's infrastructures of connection' (Couldry & Mejias, 2018, p. 337).

This new form of colonialism is digital but at the same time territorial. Most of the data extracted is controlled by companies either in the United States or in China, regardless of their origin, which means that the wealth accumulates in these regions. The data flows, thus, are global, but the practices of processing the data are embedded in arrangements of global inequality. For example, platform capitalism has given rise to new kinds of platform and data labour needed to keep the platforms' automation working and their content clean. This kind of data work is typically outsourced to the Global South, with well-known examples of content labelling in Kenya or in the Philippines (Holden & Harsh, 2024; Roberts, 2019).

In platform capitalism, thus, human beings are not only appropriated through their data but also through existing economic inequalities related to territorial and global arrangements. Despite this, introducing digital technologies and platform capitalism to the Global South has been considered progress and inclusion (Arora, 2016). For example, Facebook's Free Basics programme (Nothias, 2020) was advertised as 'connecting the unconnected', but from a business perspective, it worked to secure a first mover advantage and conquer a new customer base. Platforms, thus, are reproducing colonialistic, symbolic and material structures and power dynamics (e.g., Kumar, 2024). Another, more subtle way in which this reproduction takes place is that most social media platforms are technologies designed in the Global North and, hence, their design imposes values and worldviews dominant to those cultures (Irani et al., 2020; Olsson, *this volume*; Winner, 1980).

Being among the biggest corporate actors globally, platform companies also have political power. They have become increasingly important in policy processes, allowing them, for instance, by lobbying, to push or prevent policy changes to serve their economic interests (Cohen, 2019; Gorwa et al., 2024). Recent scholarship has discussed platform companies' power in terms of their expansion into new social areas due to their market power and digital expertise (Sharon & Gellert, 2024; van Dijck et al., 2018). AI models and applications, for example, further bolster platform companies' power and expansion across social domains (Kak et al., 2023). However, the lobbying efforts of platform companies towards global climate change, such as meeting the Paris Agreement's goals, are insignificant compared to their political capital and tend to have a 'net zero' influence over climate policy (Big Tech and Climate Policy, 2021).

Platforms Are Material

An essential aspect in understanding the environmental sustainability and responsibility of digital platforms is their materiality, not only in terms of them being the infrastructure that defines how our communication flows materialise in society but also in terms of the materials needed to build and maintain them (e.g., Haider et al., 2022; Mahnke, *this volume*). As Larkin (2013) puts it, 'Infrastructures are matter that enable the movement of other matter' (p. 329). However, they become invisible in our everyday relations and assimilate into the background, as infrastructures typically do (Star, 1999). Infrastructures are invisible yet opaque: we are used to thinking of infrastructures as something very concrete, such as the railway system or electricity network, but digital communication infrastructure escapes this logic. For the everyday user, the material nature of the digital platform infrastructure is reduced to the devices used to access the services. Larger structures, such as data centres, servers, 4G networks and cables, are out of sight, perhaps even on a different continent.

However, building these infrastructures requires land, water, energy and raw materials, including the rare minerals needed for chips and batteries (e.g., Brevini & Doctor, 2023; Brodie, 2023; Crawford, 2021; Parks et al., 2023; Pasek et al., 2023). As suggested by Kannengießer and McCurdy (2021), studies of digital media should take into account three dimensions of materials: resources, energy and e-waste. These three dimensions aim to cover the material lifecycle of digital media, from the raw materials needed to manufacture the technology to the energy needed to keep it running and, finally, to the wasteful consequences of the increasingly short lifecycle of technological devices across the infrastructure.

Several scholars have explored the environmental consequences and materiality of data centres (e.g., Brodie, 2023; Edwards et al., 2024; Hogan, 2018; Velkova, 2021) and e-waste (e.g., Bridges & Maphosa, *this volume*; Lepawsky, 2018; Parikka, 2015; Thylstrup, 2019), also highlighting the complex and global interdependencies in the ecosystem of digital technology. The material consequences of platform capitalism are more profoundly experienced in the Global South, as the zones of extraction and residual processing are concentrated in formerly colonised countries (e.g., Andreucci et al., 2023; Bridges & Maphosa, *this volume*). The platform capitalism, thus, reproduces colonial structures of violence (Thylstrup, 2019). What further complicates the picture is the increasing involvement of the platform companies in building the infrastructures of green extractivism, that is, extractivist and exploitative market dynamics of climate adaptation (Brodie, 2024; Bruna, 2023; Hogan, 2018). The platform and other big tech companies are aiming to bolster their role in the green transition by generating relationships between data centres and green electricity, particularly through discursive means.

Furthermore, the platform society has embedded logics that constantly expand its materiality. The logics of platform capitalism (Srnicek, 2017) and the surveillance economy (e.g., Zuboff, 2019) embedded in the digital platform economy generate an imperative for ever-expanding data collection (Fourcade & Healy, 2017; Hogan et al., *this volume*) and so do design solutions that encourage addictive media use (Bhargava & Velasquez, 2021; Olsson, *this volume*). This increasing use is reflected in the growing need for new data storage and calculation facilities, as well as network infrastructure, which all require material resources. As a consequence, scholars need to turn their critical gaze into material objects such as data centres, which are not only unmalleable technical systems but also 'materializations of particular power relations and imaginaries' (Edwards et al., 2024, p. 430). Yet, the language used to discuss digital technology initiatives effectively seeks to hide their materiality (Lucivero, 2020). This juxtaposition reminds us of the importance of studying the social and discursive construction of digital platform technologies.

Platforms Are Accountable

In recent years, the societal roles and responsibilities of digital platforms have become a growing topic of global discussion and increasingly also a target of regulation. Because of their infrastructural positions and accumulated power, platforms can be seen as subject to significant societal obligations (Helberger et al., 2018; Pickard, 2021). They have been increasingly called responsible for their actions in society, ranging from manipulative design and addiction (Bhargava & Velasquez, 2021) to their impact on public political discussions from elections to spreading false information (e.g., Hill, 2020; Pickard, 2021; Tucker et al., 2018).

However, formal processes of holding the platforms accountable for their actions are still only emerging, with the digital regulation acts by the European Union being the spearheading initiatives. Therefore, holding platforms responsible relies on processes of *social accountability*. Social accountability means that the powerful are answerable for their conduct and held responsible for their wrongdoings by various noninstitutional agents (Bovens, 2007). In the context of digital platforms, social accountability refers to the ways in which these agents, such as individual citizens or NGOs, publicly impose accountability on platforms. Their actions are typically discursive by nature, and the generation of public attention and debate is often a necessary requisite for holding any actor accountable.

Paradoxically, processes of social accountability often take place using the same digital media platforms that might themselves be targets of accountability demands. Digital media provide essential resources for citizens to participate in social accountability processes – from combatting organisational mischief to global climate action – to voice their critiques and demands directly at public authorities and private organisations (Chadwick, 2013; Hill, *this volume*; Ojala et al., 2019; Poupin, 2021). However, platform companies own and control the main online spaces that climate activists depend on to circulate their counter-narratives and amplify their calls for accountability (Kavada et al., 2023). Consequently, when social movements striving for environmental sustainability use digital platforms to manage their visibility and influence public opinion in order to put pressure on companies or regulators, they are forced to concede to platform power (Blum & Uldam, 2024).

Discourses that seek to hold platform companies responsible also invite them to respond. As typical to all companies, the platform companies engage in corporate communication to justify and frame their actions (Basu & Palazzo, 2008; Gillespie, 2010). Sustainability narratives are used to build legitimacy for platform companies as socially responsible actors, obfuscate the actual results of their climate response or delay climate action (Frig et al., *this volume*; Pantti & Karell, *this volume*). Climate change is a complex global problem that requires a range of actors and collaboration across various political domains. Consequently, climate governance actors and scholars alike have urged companies to play a key role in reducing the impact of climate change. While platform companies have, in recent years, shown a growing interest in environmental responsibility, at least in terms of setting climate goals and communicating about them (Arcieri, 2022), critical analyses are needed to unpack the foci, function and impact of platform sustainability discourses.

Mission and Structure of the Book

This book seeks to develop a critical understanding of the environmental responsibility and accountability of digital platforms. The 12 chapters are divided into three broad themes discussed in the previous sections: (1) unpacking environmental sustainability, (2) infrastructures and materiality and (3) discourses of accountability.

The chapters in the first section set the stage for the later chapters by building a broad understanding of the environmental impacts of digital platforms. The chapters do this by moving from a theoretical discussion of human–technology–environment relations to more concrete and empirical questions: the carbon footprint of digital platforms and digital service design and practices that drive overconsumption, as well as the ways proposed to mitigate both of them.

Chapter 1: The eco-materiality of digital platforms, by Martina Skrubbeltrang Mahnke, develops a conceptual framework to critically assess the environmental costs related to digital platforms. The framework is centred around the notion of eco-materiality, which builds on Don Ihde's theories on human–technology relations and seeks to highlight the human–technology–environment relations elemental for digital media practices. The chapter highlights the environmental materiality of digital platforms and frames them as deeply related to and embedded within the natural environment through the dimensions of resources, energy and e-waste. By doing this, the chapter sets the stage to critically understand digital technology from an environmental perspective without forgetting human agency and the related questions of global structural injustice.

The use of digital platforms, thus, is not only a matter of interaction or communicative relations but also a dynamic process that ties their use to the broader environmental and social implications of digital platforms. The author also reminds that this process and the users' ability to have agency in it varies tremendously across the world.

In *Chapter 2: The challenging task of estimating the carbon footprint of digital services*, Salla-Maaria Laaksonen and Gopika Premsankar dive into the all-but-simple practices of measuring the carbon emissions resulting from digital services. They first describe the main components that should be considered if trying to draw a complete picture: end-user devices, communication networks, and data centres, and then unpack the ways in which six major platform companies report their emissions. The chapter highlights the complexity of providing accurate measurements, resulting from the complexity of the ecosystem and lack of information from involved companies, as well as a lack of enforced standards and accountability obligations in the field. The authors conclude that in light of current information, despite the challenges and uncertainties related to the process of estimating emissions, actions to mitigate the impact are needed, and this requires ecosystem-wide collaboration.

Chapter 3: Understanding moral lapses in design: Cultural biases, beliefs, and narratives as factors of accidental evil, by Thomas Olsson, addresses the moral responsibility of online platform designers, drawing attention to certain cultural mindsets and values embedded in the design work. In particular, the chapter focuses on so-called 'accidental evil', which refers to the harmful results of design activity despite the designer's good intentions. Olsson discusses three factors contributing to the emergence of accidental evil in digital services: techno-solutionism, grand narratives and collective myopia. The chapter calls for more attention to cultural factors that may lead to moral lapses, or help designers to avoid them, and new narratives among the design community that would encourage more sustainable and just digitalisation endeavours.

Finally, *Chapter 4: Calculating, reducing, compensating, and disconnecting: Summons of digital excess*, by Minna Vigren, Tero Karppi, and Olli Pyyhtinen, discusses the concept of digital excess, which describes the abundant and superfluous flows of data resulting from digital media use. They argue that while digital excess is arguably a design problem (see also Olsson, *this volume*), it is frequently mediated as an end-user problem. In this chapter, the authors examine the ways in which users are invited to consider digital excess and the related environmental harms resulting from platform use. Using a situational analysis approach, they map a set of online materials that discuss the mitigation of individual carbon footprints and identify four central practices advocated in the materials: calculating, reducing, compensating and disconnecting. In conclusion, the authors highlight how the practices place the individual user as the actor responsible for digital excess and mitigating it, thus framing it in a recognisable framework of neoliberal individualist well-being.

The second section of the book scrutinises the *material and infrastructural dimensions* of the digital platform society. The chapters address questions related to data centre infrastructure sustainability, electronic waste and infrastructural power of the platform ecosystem over our knowledge production process, including research itself.

In *Chapter 5: The archival paradox: DNA-based data storage and the data hoarding crisis*, Mél Hogan, Deb Verhoeven and Tessa J. Brown discuss the environmental and material consequences of cloud computing and storage in data centres, and the resulting data hoarding crisis. As a solution to the increasing awareness of unsustainable data storage practices, the industry has begun exploring and developing less environmentally damaging data storage solutions. The authors deep dive into one such solution – data storage based on synthetic DNA – and explain its basic production principles as well as the discourses around the emerging technology. The authors conclude with a more critical angle, asking the big question of whether we need a shift in attitudes and data hoarding practices instead of (or in addition to) a new storage technology. Perhaps, instead of fantasies of endless data, we should embrace the graceful degradation of data and technologies.

Almost as an attempt to answer this, *Chapter 6: This photograph has never been in a data centre: An experimental intervention in pursuit of responsible research practice*, by Flora Mary Bartlett and Julia Velkova, engages in an interdisciplinary intervention to unpack and study a major paradox in critical data centre studies: that while it is well known that data centres have major material consequences for energy, landscape and labour, their critical investigation is very much reliant on the infrastructures they provide. The authors discuss their fieldwork focused on Swedish data centres and their own practices of using analogue photography as a critical methodological intervention – that is, to do critical studies on the platform infrastructure without contributing to the material networks of those platforms. The experiment required juggling but clearly shows how critical mindfulness regarding our own research practices is also very much needed – and possible.

Chapter 7: Platform opacity in the global e-waste dilemma, by Lauren E. Bridges and Vusumuzi Maphosa, turns the gaze to the global e-waste crisis and, more specifically, the role of retailer platforms in it – that is, the platform companies that act as marketplaces to connect producers and consumers. It has been well documented that the flow of e-waste tends to form from the Global North to the Global South. The authors engage with corporate reports and semi-structured interviews with retailers in the United States and Zimbabwe to reveal contrasting approaches to e-waste in the two regions, as well as their interconnectedness. Their analysis highlights how the two markets operate in different ways but still exhibit similar platform opacity in managing e-waste and communicating about it. Recycling programmes, in particular, allow platforms to evade responsibility for their role in e-waste management. To conclude, the authors propose a retailer platform responsibility approach to address the gaps in the allocation of responsibility across the tech supply chain.

Finally, the third and last section focuses on responsibility discourses circulating on and around platforms. Together, the chapters highlight the paradox of discourses discussed earlier in this introduction: the platforms are major enablers in raising awareness of sustainability issues, but, at the same time, they are increasingly targeted by discourses that demand responsibility from them, which consequently forces them to respond in their corporate discourse and action.

Chapter 8: Counter accounts and digital advocacy: How activists enforce platform companies' environmental accountability, by Markus Ojala and Mervi Pantti, offers a longitudinal exploration of how environmental non-governmental organisations hold platform companies accountable for contributing to climate change impacts. The authors study the development of accountability claims through the concept of 'counter account', a key activist tactic aiming to make harmful corporate activities visible. Platform companies have increasingly become targets of environmental activism in recent years due to their growing climate impact, public visibility and opinion power. However, only a few studies have addressed activist campaigns directed at platform companies. Examining climate activists' campaigns since 2009, the chapter identifies recurrent issues in their campaigns: climate activists' accountability claims concern platform companies' transition to renewable energy, climate policy advocacy, cooperation with fossil fuel companies and climate disinformation.

Chapter 9: Challenging plastic polluters: Public shaming campaigns on plastic pollution on social media platforms, by Steph Hill, examines narratives around the responsible handling of plastic waste. The chapter focuses on the Break Free from Plastic (BFFP) coalition's (a network of environmental organisations that cooperate on documenting and organising against plastic pollution) 'brand audit' campaigns, asking how activist groups channel social accountability efforts through the logic of social media platforms and how the campaign manages to pressure companies. The chapter argues that the campaigns demonstrate an evolution in the strategy of pressuring businesses over complex environmental issues, part of a trend in activism that is, itself, indicative of how social media platforms inform the conditions for environmental activism and discussion. Hill's chapter shows that focusing on companies that are vulnerable to public criticism on social media facilitates corporate accountability, as businesses need to safeguard their legitimacy by engaging in social responsibility activities.

Turning the gaze on the platforms' discourses, *Chapter 10: Greening the platforms or organisational bullshit? A typology of empty and misleading environmental communication by global platform companies*, by Meri Frig, Olga Dovbysh and Salla-Maaria Laaksonen, scrutinises the corporate social responsibility communication by seven platform companies in the United States, Europe, China and Russia. The authors use the theory of 'organisational bullshit', defined as empty and misleading talk, to show how companies use abstraction, distraction, reputation borrowing and stating the obvious to obscure their climate actions. The authors argue that the organisational bullshitting around environmental sustainability not only distracts stakeholders from the actual problems and needed actions but, in the case of platform companies, also helps in building narratives of futures in which the digital services they offer are depicted as an inevitable part of a more sustainable life. The authors also note that while the reporting practices differ by region, the discourses and bullshit types are surprisingly similar across the global platform industry.

Also examining platform discourses, *Chapter 11: Climate justice in Big Tech's environmental reporting*, by Mervi Pantti and Elis Karell, addresses Big Tech companies as moral agents who can be held responsible for addressing the climate emergency. Drawing on the literature on platform power and climate justice, the chapter argues that the moral responsibility of counteracting the uneven impacts of climate change can be assigned to platform companies because of their powerful position. The concept of climate justice highlights the unequal impacts of climate change as well as inequalities regarding actors' different responsibilities and capabilities in addressing climate change. By focusing on human rights, climate justice introduces alternative perspectives alongside the dominant technology- and market-oriented discourse on climate change. The chapter examines the ways in which climate justice is articulated in Big Tech's environmental reporting. It shows that despite ambitious climate goals and adoption of climate justice terminology, platform companies pay limited attention to climate justice issues in Global South countries, and their corporate activities in Global South reflect the economic and technological power of the companies – large investments in renewable energy, extensive partnerships and technological or data-based solutions.

The volume concludes with an afterword by Simon Cottle. He offers profound thoughts on digital platforms and environmental responsibility, situating them within today's 'world-in-crisis' (Cottle, 2025), a world of systemic and entangled crises, including climate change, biodiversity loss, global pandemics and water and energy insecurity, that pose existential threats to human and nonhuman life. He convincingly argues that climate change needs to be explained and responded to in relation to a world in which the age of fossil-fuelled extractivism and rampant ecological exploitation – a worldview based on economic growth and human exclusivism – must come to an end. Enormously powerful Big Tech corporations, he writes, shape corporate and political activity, professional and cultural practices and individual identities. While they offer global networks of connectivity and give channels for voices aiming for ecological and societal change, in the end, they are built on a worldview that does not question continuous economic growth, social injustice and ever-increasing individualistic consumption.

Where Should We Go From Here?

Digital media platforms have gained infrastructural power in our society as mediators and enablers of our social lives, public discussions and political action. As the chapters in this book highlight, they are also deeply intertwined with questions of environmental sustainability, both by providing a space where discussions related to the climate crisis are circulating and as drivers of the crisis themselves. The irresistible pull of digitalisation and the ever-increasing consumption of digital content that platforms invite us to are simultaneously producing a world that requires increasingly massive material infrastructures, data storage and more energy. They stimulate us to join a digital world of abundance and excess (Olsson et al., 2023; Vigren et al., *this volume*), but it has become increasingly clear that our planetary resources cannot sustain such a world (Istrate et al., 2024). Acknowledging this invites a shift in thinking about how environmental responsibility and accountability are constructed around digital platforms.

Following the empirical insights and arguments provided in the chapters of this book, we argue that the world urgently needs growing awareness of the environmental impacts of digital technology and a broader discussion about the platform companies' responsibility for it. We should actively seek to decouple the prominent perception of digitalisation as an environmentally sustainable development and instead make the understanding of the environmental impact of digital technologies and the platform industry behind them common knowledge at all levels, from users to policymakers. This also requires a critical stance towards the techno-optimistic narratives that are being pushed by the digital technology companies, and in the long term, also a broader rethinking of the capitalistic and neo-colonialistic structures of the platform economy.

Sustainability research has identified three ways to reduce carbon emissions: reducing consumption, improving energy efficiency and modal shifts from one type of consumption pattern to one that uses fewer resources (Lettenmeier et al., 2019). These three methods should also be at the heart of our efforts to address the environmental impacts of digital media consumption, similar to other forms of consumption. Research on the sustainability transition often focuses on energy efficiency and reducing consumption rather than on finding alternative, more ecologically sustainable consumption patterns (modal shifts). Striving towards such shifts requires a better understanding of the ways in which digital services cause environmental harm (e.g., Bridges & Maphosa, *this volume*; Hogan et al., *this volume*; Laaksonen & Premsankar, *this volume*) and how digital services are designed with accidentally harmful environmental impacts (Olsson, *this volume*). Eventually, it also requires political will to regulate digital activities and to seek collective ways to reduce the footprint resulting from our digital lives, instead of pushing the responsibility to the individual (Vigren et al., *this volume*). As several authors in this book point out, this responsibility cannot be shifted to consumers, but the critical gaze should target the entire system and the underlying logics of platform capitalism (see also Gritsenko et al., 2024). Nonetheless, with this book, we contribute to the necessary first step: to generate awareness about the environmental impact of digital platform technology.

References

Alaimo, C., & Kallinikos, J. (2024). *Data rules: Reinventing the market economy*. The MIT Press.

Alcott, B. (2005). Jevons' paradox. *Ecological Economics*, *54*(1), 9–21.

Andreucci, D., López, G. G., Radhuber, I. M., Conde, M., Voskoboynik, D. M., Farrugia, J. D., & Zografos, C. (2023). The coloniality of green extractivism: Unearthing decarbonisation by dispossession through the case of nickel. *Political Geography*, *107*, 102997. https://doi.org/10.1016/j.polgeo.2023.102997

Arcieri, K. (2022). *Path to net-zero: Datacenter demands push Amazon, Big Tech toward renewables*. S&P Global Market Intelligence. https://www.spglobal.com/marketintelligence/en/news-insights/latest-news-headlines/path-to-net-zero-data-center-demands-push-amazon-big-tech-toward-renewables-72752727

Arora, P. (2016). Bottom of the data pyramid: Big Data and the Global South. *International Journal of Communication*, *10*(0), 19. https://ijoc.org/index.php/ijoc/article/view/4297

Basu, K., & Palazzo, G. (2008). Corporate social responsibility: A process model of sensemaking. *Academy of Management Review*, *33*(1), 1–43. https://doi.org/10.5465/amr.2008.27745504

Bhargava, V. R., & Velasquez, M. (2021). Ethics of the attention economy: The problem of social media addiction. *Business Ethics Quarterly*, *31*(3), 321–359. https://doi.org/10.1017/beq.2020.32

Big Tech and Climate Policy. (2021, January). An InfluenceMap report. https://influencemap.org/report/Big-Tech-and-Climate-Policy-afb476c56f217ea0ab351d79096df04a

Blum, I., & Uldam, J. (2024). Faking, optimising and conceding to power: Social movement understandings of social media power. *New Media & Society*. Advance online publication. https://doi.org/10.1177/14614448241266769

Bovens, M. (2007). Analysing and assessing accountability: A conceptual framework. *European Law Journal*, *13*, 447–468. https://doi.org/10.1111/j.1468-0386.2007.00378.x

Brevini, B., & Doctor, D. (2023). Carbon capitalism, communication, and artificial intelligence: Placing the climate emergency center stage. In A. López, A. Ivakhiv, S. Rust, M. Tola, A. Y. Chang, & K. Chu (Eds.), *The Routledge handbook of eco-media studies* (pp. 171–178). Routledge. https://doi.org/10.4324/9781003176497-21

Brevini, B., & Swiatek, L. (2020). *Amazon: Understanding a global communication giant*. Routledge.

Brodie, P. (2023). Data infrastructure studies on an unequal planet. *Big Data and Society*, *10*(1). https://doi.org/10.1177/20539517231182402

Brodie, P. (2024). Smarter, greener extractivism: Digital infrastructures and the harnessing of new resources. *Information, Communication & Society*, *28*(6), 1061–1080. https://doi.org/10.1080/1369118X.2024.2341013

Bruna, N. (2023). *The rise of green extractivism: Extractivism, rural livelihoods and accumulation in a climate-smart world*. Routledge.

Buchanan, J. M., & Stubblebine, W. C. (1962). Externality. *Economica*, *29*, 371–384.

Chadwick, A. (2013). *The hybrid media system. Politics and power*. Oxford University Press.

Cohen, J. (2019). *Between truth and power: The legal constructions of informational capitalism*. Oxford University Press.

Cottle, S. (Ed.). (2025). *Communicating a world-in-crisis*. Peter Lang.

Couldry, N., & Mejias, U. A. (2018). Data colonialism: Rethinking big data's relation to the contemporary subject. *Television & New Media*, *20*(4), 336–349. https://doi.org/10.1177/1527476418796632

Crawford, K. (2021). *Atlas of AI: Power, politics, and the planetary costs of artificial intelligence*. Yale University Press. https://doi.org/10.12987/9780300252392

Edwards, D., Cooper, Z. G. T., & Hogan, M. (2024). The making of critical data center studies. *Convergence*, *31*(2), 429–446. https://doi.org/10.1177/13548565231224157

Fourcade, M., & Healy, K. (2017). Seeing like a market. *Socio-Economic Review, 15*(1), 9–29. https://doi.org/10.1093/ser/mww033

Freitag, C., Berners-Lee, M., Widdicks, K., Knowles, B., Blair, G. S., & Friday, A. (2021). The real climate and transformative impact of ICT: A critique of estimates, trends, and regulations. *Patterns, 2*(9), 100340. https://doi.org/10.1016/J.PATTER.2021.100340

Gillespie, T. (2010). The politics of "platforms." *New Media & Society, 12*(3), 347–364.

Gorwa, R., Lechowski, G., & Schneiß, D. (2024). Platform lobbying: Policy influence strategies and the EU's Digital Services Act. *Internet Policy Review, 13*(2), 1–26.

Gritsenko, D., Aaen, J., & Flyvbjerg, B. (2024). Rethinking digitalization and climate: Don't predict, mitigate. *npj Climate Action, 3*(1), 1–7. https://doi.org/10.1038/s44168-024-00127-z

Haider, J., Rödl, M., & Joosse, S. (2022). Algorithmically embodied emissions: The environmental harm of everyday life information in digital culture. *Information Research, 27*. https://doi.org/10.47989/colis2224

Helberger, N. (2020). The political power of platforms: How current attempts to regulate misinformation amplify opinion power. *Digital Journalism, 8*(6), 842–854. https://doi.org/10.1080/21670811.2020.1773888

Helberger, N., Pierson, J., & Poell, T. (2018). Governing online platforms: From contested to cooperative responsibility. *The Information Society, 34*(1), 1–14. https://doi.org/10.1080/01972243.2017.1391913

Hill, S. (2020). Politics and corporate content: Situating corporate strategic communication between marketing and activism. *International Journal of Strategic Communication, 14*(5), 317–329. https://doi.org/10.1080/1553118X.2020.1817029

Hogan, M. (2018). Big data ecologies. *ephemera, 18*(3), 631–657.

Holden, K., & Harsh, M. (2024). On pipelines, readiness and annotative labour: Political geographies of AI and data infrastructures in Africa. *Political Geography, 113*, 103150. https://doi.org/10.1016/j.polgeo.2024.103150

Irani, I., Vertesi, J., Dourish, P., Philip, K., & Grinter, R. E. (2010). Postcolonial computing: A lens on design and development. In *Proceedings of the SIGCHI conference on human factors in computing systems (CHI '10)* (pp. 1311–1320). Association for Computing Machinery. https://doi.org/10.1145/1753326.1753522

Istrate, R., Tulus, V., Grass, R. N., Vanbever, L., Stark, W. J., & Guillén-Gosálbez, G. (2024). The environmental sustainability of digital content consumption. *Nature Communications, 15*(1), 1–11. https://doi.org/10.1038/s41467-024-47621-w

Kak, A., Myers West, S., & Whittaker, M. (2023, December 5). *Make no mistake—AI is owned by Big Tech.* MIT Technology Review. https://www.technologyreview.com/2023/12/05/1084393/make-no-mistake-ai-is-owned-by-big-tech

Kannengießer, S., & McCurdy, P. (2021). Mediatization and the absence of the environment. *Communication Theory, 31*(4), 911–931. https://doi.org/10.1093/ct/qtaa009

Kavada, A., Askanius, T., Kaun, A., Mattoni, A., & Uldam, J. (2023, February 14). *Tying up Goliath: Activist strategies for confronting and harnessing digital power.* Transnational Institute. https://www.tni.org/en/article/tying-up-goliath

Khanal, S., Zhang, H., & Taeihagh, A. (2024). Why and how is the power of Big Tech increasing in the policy process? The case of generative AI. *Policy and Society*. https://doi.org/10.1093/polsoc/puae012

Kumar, S. (2024, July 17). *The digital and the postcolonial.* Oxford Research Encyclopedia of Communication. https://doi.org/10.1093/acrefore/9780190228613.013.1280

Kuntsman, A. (2020). From digital solutionism to materialist accountability: The urgency of new interventions. *Journal of Environmental Media, 1*(1), 15–19. https://doi.org/10.1386/jem_00002_1

Kuntsman, A., & Rattle, I. (2019). Towards a paradigmatic shift in sustainability studies: A systematic review of peer reviewed literature and future agenda setting to consider environmental (un)sustainability of digital communication. *Environmental Communication, 13*(5), 567–581. https://doi.org/10.1080/17524032.2019.1596144

Laaksonen, S. M., Koivula, M., & Villi, M. (2024). Mediated by the giants: Tracing practices, discourses, and mediators of platform isomorphism in a media organization. *New Media & Society, 26*(8), 4317–4335. https://doi.org/10.1177/14614448221122220

Larkin, B. (2013). The politics and poetics of infrastructure. *Annual Review of Anthropology, 42*(1), 327–343. https://doi.org/10.1146/annurev-anthro-092412-155522

Lehdonvirta, V. (2022). *Cloud empires: How digital platforms are overtaking the state and how we can regain control.* MIT Press.

Lepawsky, J. (2018). *Reassembling rubbish: Worlding electronic waste.* The MIT Press.

Lettenmeier, M., Akenji, L., Toivio, V., Koide, R., & Amellina, A. (2019). *1.5 degree lifestyles: Targets and options for reducing lifestyle carbon footprints.* Institute for Global Environmental Strategies. https://www.iges.or.jp/en/pub/15-degrees-lifestyles-2019/en

Lucivero, F. (2020). Big data, big waste? A reflection on the environmental sustainability of big data initiatives. *Science and Engineering Ethics, 26*, 1009–1030. https://doi.org/10.1007/s11948-019-00171-7

Mejias, U. A., & Couldry, N. (2024). *Data grab: The new colonialism of big tech and how to fight back.* University of Chicago Press.

Moore, J. W. (2015). *Capitalism in the web of life: Ecology and the accumulation of capital.* Verso.

Morozov, E. (2014). *To save everything, click here: The folly of technological solutionism.* Public Affairs.

Narayan, D. (2022). Platform capitalism and cloud infrastructure: Theorizing a hyper-scalable computing regime. *Environment and Planning A, 54*(5), 911–929. https://doi.org/10.1177/0308518X221094028

Nieborg, D., Poell, T., Caplan, R., & van Dijck, J. (2024). Introduction to the special issue on locating and theorising platform power. *Internet Policy Review, 13*(2), 1–17. https://doi.org/10.14763/2024.2.1781

Nothias, T. (2020). Access granted: Facebook's free basics in Africa. *Media, Culture & Society, 42*(3), 329–348. https://doi.org/10.1177/0163443719890530

Ojala, M., Pantti, M., & Laaksonen, S. M. (2019). Networked publics as agents of accountability: Online interactions between citizens, the media and immigration officials during the European refugee crisis. *New Media & Society, 21*(2), 279–297. https://doi.org/10.1177/1461444818794592

Olsson, T., Pyyhtinen, O., Laaksonen, S.-M., Vigren, M., Ylipulli, J., Rantasila, A., & Sawhney, N. (2023). A user-centered lens into digital excess: Exploring the superfluity and environmental burden of the digital world. In *Proceedings of the LIMIT ninth computing within limits 2023*. https://doi.org/10.21428/bf6fb269.a5916a92

Parikka, J. (2015). *A geology of media*. University of Minnesota Press.

Parks, L., Velkova, J., & De Ridder, S. (Eds.). (2023). *Media backends: Digital infrastructures and sociotechnical relations*. University of Illinois Press.

Pasek, A., Vaughan, H., & Starosielski, N. (2023). The world wide web of carbon: Toward a relational footprinting of information and communications technology's climate impacts. *Big Data and Society*, *10*(1). https://doi.org/10.1177/20539517231158994

Pickard, V. (2021). A new social contract for platforms. In M. Moore & D. Tambini (Eds.), *Regulating Big Tech: Policy responses to digital dominance*. Oxford Academic. https://doi.org/10.1093/oso/9780197616093.003.0018

Plantin, J. C., Lagoze, C., Edwards, P. N., & Sandvig, C. (2018). Infrastructure studies meet platform studies in *the* age of Google and Facebook. *New Media & Society*, *20*(1), 293–310. https://doi.org/10.1177/1461444816661553

Poupin, P. (2021). Social media and state repression: The case of VKontakte and the anti-garbage protest in Shies, in Far Northern Russia. *First Monday*, *26*(5). https://doi.org/10.5210/fm.v26i5.11711

Roberts, S. T. (2019). *Behind the screen: Content moderation in the shadows of social media*. Yale University Press.

Sharon, T., & Gellert, R. (2024). Regulating Big Tech expansionism? Sphere transgressions and the limits of Europe's digital regulatory strategy. *Information, Communication & Society*, *27*(15), 2651–2668. https://doi.org/10.1080/1369118X.2023.2246526

Srnicek, N. (2017). *Platform capitalism*. Polity Press.

Star, S. L. (1999). The ethnography of infrastructure. *American Behavioral Scientist*, *43*(3), 377–391.

Thylstrup, N. B. (2019). Data out of place: Toxic traces and the politics of recycling. *Big Data & Society*, *6*(2). https://doi.org/10.1177/2053951719875479

Tucker, J. A., Guess, A., Barberá, P., Vaccari, C., Siegel, A., Sanovich, S., Stukal, D., & Nyhan, B. (2018). *Social media, political polarization, and political disinformation: A review of the scientific literature*. A Hewlett Foundation Report. https://doi.org/10.2139/ssrn.3144139

van Dijck, J. (2021). Seeing the forest for the trees: Visualizing platformization and its governance. *New Media & Society*, *23*(9), 2801–2819. https://doi.org/10.1177/1461444820940293

van Dijck, J., & Poell, T. (2015). Social media and the transformation of public space. *Social Media + Society*, *1*(2). https://doi.org/10.1177/2056305115622482

van Dijck, J., Poell, T., & de Waal, M. (2018). *The platform society: Public values in a connective world*. Oxford University Press.

Velkova, J. (2021). Thermopolitics of data: Cloud infrastructures and energy futures. *Cultural Studies*, *35*(4–5), 663–683. https://doi.org/10.1080/09502386.2021.1895243

Willson, M. (2014). The politics of social filtering. *Convergence: The International Journal of Research into New Media Technologies*, *20*(2), 218–232. https://doi.org/10.1177/1354856513479761

Winner, D. (1980). Do artifacts have politics? *Dædalus*, *109*(1), 121–136.

World Bank & ITU. (2024). *Measuring the emissions & energy footprint of the ICT sector: Implications for climate action*. https://www.itu.int/hub/publication/d-ind-clim-2023-01/

Zuboff, S. (2019). *The age of surveillance capitalism: The fight for a human future at the new frontier of power*. Public Affairs.

Section 1

Unpacking Environmental Sustainability

Chapter 1

The Eco-Materiality of Digital Platforms

Martina Skrubbeltrang Mahnke

Roskilde University, Denmark

Abstract

Digital platforms, such as Facebook, Instagram and TikTok, provided by big tech companies, have become central infrastructures in contemporary societies. However, they are also considered to be among the largest hidden contributors to CO_2 emissions. Given the severity of the climate crisis, this highlights the need for a critical assessment of how to study digital platforms with regard to their environmental costs and related structures of global injustice. The aim of this chapter is to develop a conceptual framework centred on the notion of eco-materiality, encompassing infrastructural dimensions such as energy use, resources and e-waste, as well as socio-contextual aspects such as user perceptions and cultural context. Examining digital platforms through an eco-material lens sheds light on the human–technology–environment relations, showing that digital practices are deeply interwoven with the consumption of environmental resources and are entangled in global injustices. The developed relational framework provides a valuable starting point for critical empirical investigations into digital platforms and their environmental and social embeddedness. It also invites researchers to rethink current theoretical conceptualisations of digital platforms by incorporating socio-environmental aspects, while maintaining a focus on human agency and the critical role that humans play in these relations.

Keywords: Digital platforms; environment; materiality; platform use; human–technology relations; eco-materiality; environmental costs; human agency

Platforms and the Planet, 21–36

doi:10.1108/978-1-83662-172-020261002

Introduction

Digital platforms, such as TikTok, Facebook, Instagram, Netflix and YouTube, operated by big tech companies, have become central societal infrastructure(s). They provide essential communication services, ranging from education to entertainment. However, they also rank highest among the largest concealed contributors to current CO_2 emissions (Sharma & Dash, 2022). Their perceived indispensability incurs significant environmental costs, calling for a thorough and critical examination of their environmental impact. Recent public discourse has begun addressing the environmental costs of digital platforms and their uneven distribution as a major concern; however, within the tech industry, it is often downplayed or regarded as an issue to address later. For example, in response to a LinkedIn post on ChatGPT's immense energy consumption, a data analytics professional commented: 'This is truly the least of our concerns'. Yet with the rapid growth and implementation of generative artificial intelligence (AI) in the digital sphere, the environmental footprint of digital platforms is expected to grow exponentially (Saenko, 2023), making it an urgent societal issue. Even Sam Altman, OpenAI's CEO, stated that securing the vast amounts of energy needed to sustain digital platforms poses a significant challenge for the tech industry and, by extension, contemporary societies (Dastin, 2024). While tech industry leaders have started to acknowledge the energy challenges, they rarely address how these costs disproportionately impact communities that are already vulnerable to and burdened by the climate crisis (Kneese & Young, 2024; also Pantti & Karell, *this volume*).

The rising environmental costs of digital platforms often remain hidden behind their intangible nature. During the early stages of societal digitisation, the phrase 'think before you print' became a well-known reminder of the high environmental impact of print media, promoting the notion of digital technologies as a more sustainable choice. Recently, a few studies have started to point out that digital technologies rely heavily on natural resources, including energy consumption, data storage and the extraction of minerals (Haider et al., 2022). However, the commonly perceived immateriality of digital technologies fosters a widespread belief in their minimal environmental impact. This perception persists despite estimates from over a decade ago, in 2011, suggesting that digital communication activities produce as much CO_2 as the aviation industry (Jones, 2018; Raghavan & Ma, 2011; Vestberg et al., 2014). While energy efficiency has increased in recent years, the growing number of users has outweighed these advancements, resulting in an overall increase in CO_2 emissions from digital technologies (Malmodin et al., 2024). Currently, the environmental impact of digital platforms is not only overlooked but so is the disproportionate burden that digitalisation processes place on communities most affected by the climate crisis. Today, the disconnect between the digital sphere and its environmental and social consequences is reflected in the general assumption that data and digital technologies are primarily solutions to climate change rather than contributors to it (Wang et al., 2023).

The widely recognised public launch of ChatGPT in November 2022 sparked new public debate regarding the balance between technological benefits and their environmental costs. Although public awareness of the immense use of environmental resources is growing, precise knowledge about these costs remains limited. Calculation models are complex, particularly in accounting for factors such as energy consumption, manufacturing and usage (Batmunkh, 2022; Laaksonen & Premsankar, *this volume*). However, researchers generally agree that the global scale and ubiquity of digital platforms and AI-driven forms of communication present unprecedented environmental challenges (Hao, 2019). In response to rising public criticism, (big) tech actors have started 'greening' their technologies by transitioning to renewable energy sources and promoting sustainable agendas. Despite these 'green' initiatives, digital societies are witnessing a steep rise in emissions from digital technologies (Ferreboeuf et al., 2021). Microsoft, as one of the key organisational actors, reported a 30% increase in its emissions, mainly due to increased user demand and the rising energy costs of AI systems (Kimball, 2024). At the same time, greening efforts, often concentrated in the Global North, frequently shift environmental burdens to the Global South, thereby intensifying existing inequalities (Natarajan, 2021).

In light of the severity of the climate crisis, this chapter raises critical questions regarding how to study and conceptualise digital platforms in relation to their environmental costs and associated social injustices. The aim is to develop a comprehensive conceptual framework that accounts for the material aspects of digital platforms, their sociocultural aspects and human agency. Emphasising the human–technology link is crucial, as it is ultimately people that bear the consequences of climate change and therefore share the responsibility for addressing it (Coeckelberg, 2021). While ideally this responsibility is shared, the capacity to act differs greatly due to the uneven distribution of power and resources between the Global North and the Global South (Rice, 2007). Drawing on Ihde's (1990) human–technology relations and Kannengießer and McCurdy's (2021) work on the environmental aspects of digital technologies, the overall aim of this chapter is to make the human-platform-environment nexus both theoretically and empirically tangible. Additionally, it discusses human agency as a crucial dimension, exploring how different actors not only relate differently to digital platforms but also possess different forms of agency in shaping the environmental consequences. It therefore aims to provide a solid foundation for advancing theoretical approaches and fostering critical analyses.

The Environment: A Blind Spot in Digital Platform Research

Recent studies on digital platforms have extensively mapped and analysed the infrastructural and platformised power inherent in and constitutive of digital platforms (Flyverbom et al., 2019; Helberger et al., 2018; Lai & Flensburg, 2023; Pickard, 2021). Studies have further argued that underlying capitalistic structures primarily drive the development of such platforms (Roos et al., 2016).

Questions regarding the environmental and social implications of digital platforms, however, receive relatively little attention and largely remain a blind spot in current research. Only a few studies have explicitly examined the materiality of digital platforms, investigating in greater depth how these technologies are 'made' and 'trained' based on the exploitation of natural resources (Crawford, 2021). Global structural dynamics further reinforce environmental injustices, as the benefits from digital innovation processes are unevenly distributed in the Global North and the Global South. More recent research at the intersection of environmental communication and sustainability studies has started to focus on digital platforms in this relation, reporting that the damage inflicted on the natural world generally remains unnoticed, uncommunicated and under-theorised (Kuntsman & Rattle, 2019). A major point of critique is the dominant narrative of the great benefits of digital platforms, which tends to neglect and downplay their extensive environmental and social costs (Borning et al., 2018). On a broader level, researchers across disciplines agree that the focus on the benefits of digital platforms often marginalises sustainable approaches, which can be defined as 'the balance between what the Earth can give to support human activity and what the Earth can safely re-absorb from those activities' (Maxwell, 2014, p. 36).

Another aspect of the environmental blind spot in digital platform research emerges when focusing on communication and media studies. While environmental issues are studied broadly, the focus often centres on media content, such as how journalists report on the climate crisis (Schäfer & Schlichting, 2014) or how users engage in repairing and recycling their digital devices (Kannengießer, 2018). However, a few media scholars have started highlighting the lack of critical research examining digital technologies as composed of raw materials and resources extracted *from* the environment, rather than as providers of content *about* the environment (López et al., 2024).

A common critique of studies on digital platforms is their tendency to overlook how human conditions, such as usage, sociocultural factors and human agency co-shape these platforms (Mathieu & Pruulmann-Vengerfeldt, 2020). While research has shed light on how citizens tackle rising environmental challenges in relation to their consumption of digital devices (Kannengießer, 2020; Stephansen & Treré, 2020), a significant gap remains in studies that make the environmental materiality of digital platforms as well as the diverse roles and varying degrees of agency of human users visible, actionable and meaningful. Recognising the central role that digital platforms play in contemporary societies, a key question is how to conceptualise the intricate relationships between human agency, digital platforms and the environment. As a preliminary answer, this chapter proposes an eco-material human-centred framework that expands Ihde's (1990) conceptualisations of human–technology relations and connects them to the environmental materiality of digital platforms in the form of energy use, resource consumption and e-waste (Kannengießer & McCurdy, 2021).

Digital Platforms as Human–Technology Relations

A widely referenced definition is van Dijck's (2013) understanding of digital platforms as a 'set of relations that constantly needs to be performed' (p. 26). This definition addresses platform-inherent power relations and points to the ongoing friction between individual users' goals and platform providers. While platform providers typically pursue profit-oriented goals, users engage with platforms to fulfil their own needs. As stated above, researchers generally agree that economic interests prevail in digital platforms and that communicative acts, such as shares and likes, are mainly implemented to support this interest rather than foster social relations (Plantin et al., 2016). From this perspective, platforms are examined as technological structures that use big data and automated analysis to create personalised user experiences. While economic structures may dominate, user studies have shown that users are still able, albeit in a limited way, to exert agency, circumvent and even counter technological structures in different, often creative, ways (Bonini & Treré, 2024). This shows that despite structural dominance, users are not merely consumers but actively engage with digital platforms, emphasising the dynamic nature of human–technology relations (Mahnke et al., 2024).

As digital platforms have become embedded in everyday life, an alternative way of describing the user-platform relationship is through Don Ihde's (1990) notion of intentionality (Verbeek, 2001). Intentionality contrasts with the goal-conscious activities of human beings and views users and technology as inevitability linked. This implies that the relationship between users and platforms is not merely instrumental but mutually shaping. Digital platforms not only shape human users and perceptions but are equally shaped by them. By emphasising the relational dynamics, Ihde's concept helps illuminate the deeper socio-technical entanglements that constitute digital platforms. The philosophy of technology has a long tradition of exploring the complex dynamics that constitute human–technology relationships, particularly regarding how human conditions, such as belief systems, norms, values and expectations, shape technologies (Cressman, 2022; Kerr et al., 2020). Even long before the digital era, philosophers contemplated how technology shapes human experience(s) of and within the world. When exploring the question of how to conceptualise the relationship between human agency, digital platforms and the environment, such philosophical inquiries serve as a valuable starting point due to their focus on the human–technology relationship.

Ihde's (1990) phenomenological approach supports overcoming the classic distinction between object and subject, focusing instead on their interrelatedness. According to Ihde, object and subject cannot be viewed independently but only in relation to each other. The 'I', the subject, and the 'world', the contextual surroundings, do not exist as separate entities but are always interconnected. In Ihde's understanding, humans do not experience the world directly; rather, the world is revealed to them through mediating factors. 'Technology', as a mediating factor, plays an important role in this revealing process, as it mediates human experience. This mediated relationship shows how technologies shape the ways humans perceive and engage with their surroundings. In his work, Ihde (1990) investigates how the 'I' (the subject), 'technology' and the 'world' (contextual surroundings) are related to one another, distinguishing between four human–technology relations (see Fig. 1.1): (1) embodied relations, (2) hermeneutic relations, (3) alterity relations and (4) background relations.

In the first, *embodied relations*, technology becomes an extension of the body; in other words, technology emerges. A typical example is spectacles or prosthetics, where the 'I' and the 'technology' are closely intertwined. In these cases, the technology becomes almost invisible or secondary to the human user. For example, when wearing glasses, users do not focus on the glasses themselves, but on what they can see through them. Technology, in the form of the glasses, becomes closely bound to the user, who then perceives the 'world' or the contextual surroundings, more clearly. Similarly, prosthetics, another example mentioned by Ihde, are not merely external tools but extensions of the human body, allowing certain interactions that would otherwise be impossible. In embodied relations, the boundaries between the 'I', the subject and the 'technology' seem to dissolve, creating an experience in which the technology and the human user are perceived as working together as a whole.

In the second human–technology relation, which Ihde termed *hermeneutic relations*, technology provides a means of interpreting the world, becoming performative in the process. Examples include a thermometer that measures the temperature or a map that represents a certain part of the world. In this relation, the 'I', the subject, becomes more distant, while the 'world' and the 'technology' move closer together. The 'I' then assumes the role of interpreter, as the technology has become a representation of the world. In the case of the thermometer, the human needs to interpret the numerical value displayed. Thus, the technology, in the form of a thermometer, becomes a tool through which reality is experienced and made tangible. The role of the human shifts from direct interaction with the world to interpreting the representations provided by the technology, which serves as a lens through which the world is perceived.

In the third relation, termed *alterity relations*, technology is experienced as interactive and distinct from the human user. It is not merely an extension of the body but is experienced as independent and autonomous. This type of interaction can occur, for example, when engaging with a chatbot. In alterity

Embodied relations: (I-technology) -> world (technology is emerged)

Hermeneutic relations: I -> (technology-world) (technology is performative)

Alterity relations: I -> technology (-world) (technology is interactive)

Background relations: I -> technology (-world) (technology is contextual)

Fig. 1.1. Don Ihde's Technology Relations. *Source:* Own illustration.

relations, technology is more than just a tool and may be experienced as having agency. Technology is thus no longer seen solely as an extension of the body, as in embodied relations, or as a means of interpreting the world, as in hermeneutic relations, but is typically encountered in the form of an 'other'. This sense of otherness arises because the technology interacts autonomously and interdependently. Oftentimes, this begins to blur subject-object boundaries, as the technology may be perceived as another 'I' rather than as a type of technology.

For example, recent research on ChatGPT has shown that users experience generative AI technology as a 'happy helper' or 'conversation partner' rather than as a technology (Mahnke & Bagger, 2024). As a result of this sense of otherness, in alterity relations, the 'I' can, at times, have strong emotional experiences and/or feel as if they are engaging in a kind of social relationship with the technology. In alterity relations, the focus is on the I-technology relation, which causes the 'world' to move into the background. This shift shows how the technology itself becomes central to the user experience, while its role as a means through which the world is experienced becomes secondary.

In the fourth and final type, *background relations*, the technology operates in the periphery, shaping the context of an experience rather than being directly in focus. An example of this is noise-cancelling systems, which are designed to create a quieter experience without demanding constant user interaction. In background relations, the technology typically only becomes visible when it malfunctions or needs adjustment. Here, the focus also shifts towards the I-technology relationship, while the world as contextual surroundings becomes less tangible. In both alterity and background relations, the technology becomes more prominent and independent, either in the form of another, as in alterity relations, or as an influencing contextual factor, as in background relations.

When exploring digital platforms through the lens of Ihde's human–technology relations, one apparent reading that emerges is that as digital platform technology evolves into more intelligent systems, it starts to separate the I, the human user, from directly experiencing the world, which also includes the use of environmental resources. Although this separation exists in all four types of relations, it becomes particularly pronounced in alterity and background relations. In both, the emphasis shifts to the I-technology relation, albeit in different ways, which obscures the physical materiality of digital platforms and deepens the disconnect from their ecological implications. The following section will therefore focus on these two relations and develop the analysis around them.

As digital platforms become more integrated into everyday life, users engage primarily with the platform's interface rather than the technological infrastructure that underpins it. Consequently, the environmental impact of digital platforms, such as the use of energy or resources, is perceived – if at all – as distant and abstract. In alterity relations, technologies tend to take on the role of what Gunkel (2012) has termed a 'communicative other'. As stated above, in this relation, technologies are not experienced as tools or material infrastructures but rather as active communicative agents that participate in human communication processes. Human–technology relations, in which digital platforms are perceived as communicative entities enabling human-like conversations, contribute to a growing disconnect between the user and the environmental realities of these platforms. In other words, digital platforms are becoming environmentally unrelatable. This disconnect obscures the environmental materiality of digital platforms, making it challenging for users to relate to it. As a result, users typically interact with the platform as if it operates independently of any environmental implications. The environmental unrelatability of the platforms further deepens global social injustices, as marginalised communities affected by the climate crises remain unseen (Rosignoli et al., 2023).

The disconnect between the use of platforms and their environmental implications raises several critical questions. Within background relations, the consumption of material resources largely goes unnoticed, as users are typically unaware of the vast amounts of energy and resources required to run digital infrastructures. This creates a gap in which the relationship between digital platforms and the natural world remains hidden and inaccessible to human experience. Simultaneously, platforms in alterity relations bring users closer to an experience of technology as human-like, further distancing them from recognising the environmental impact. When users experience technology as human-like, the emphasis shifts to the communicative relationship between the user and the technology (Schwartz & Mahnke, 2021). To reiterate, the invisibility, intangibility and unrelatability of digital platforms in an environmental sense contribute to users' disconnect from perceiving any environmental and social implications. When exploring digital platforms through Ihde's lens of human–technology relations, it becomes clear that addressing the environmental implications of digital platforms requires rethinking human–technology relations within an environmental context.

The Environmental Materiality of Digital Platforms

When seeking to understand the environmental materiality of digital platforms, it is essential to frame them as deeply related to and embedded within the natural environment. In this context, Ihde's (1990) conceptualisation of the 'world' as a contextual environment requires an extension, which includes the environment as a material reality, positioning it as an integral part of this 'world'. Incorporating environmental materiality into Ihde's concept of the 'world' allows for a more comprehensive analysis of digital platforms and their environmental implications. In this regard, Kannengießer and McCurdy (2021) advocate for a perspective of the environment that centres on three dimensions: resources, energy and e-waste. The first dimension, resources, emphasises the extraction of raw materials required for manufacturing digital technologies and infrastructures. As such, resources are not only limited but are often sourced through environmentally damaging processes and ethically questionable practices (Lucivero, 2019). The second dimension, energy, points to the extensive amounts of energy needed to power and sustain digital platforms, particularly in relation to energy-intensive data centre activities and AI training platforms. As the demand for and use of AI in digital platforms increases, energy consumption will continue to rise, posing significant challenges for digital societies (Crawford, 2021). The third dimension, e-waste, addresses the rapid turnover rate and disposability of digital technologies, which often have rather short lifecycles. As digital technologies are rapidly evolving, devices quickly become outdated and obsolete, contributing to the significant accumulation of digital and electronic waste (Bridges & Maphose, *this volume*; Kannengießer, 2020).

Exemplifying the environmental materiality of digital platforms through the dimensions of resources, energy and e-waste shifts the focus to the material realities of these platforms. This approach makes the use of environmental resources in digital platforms tangible and provides valuable entry points for critical empirical analysis. Drawing on Gabrys' work (2011) and Kannengießer and McCurdy (2021) further propose understanding the eco-materiality of digital platforms as a process. In this view, the environmental materiality of digital platforms is not primarily a property or consequence but a fundamental component embedded within the human–technology relations that constitute these platforms. Understanding the environmental materiality of digital platforms as a process rather than as a sole mechanistic capacity is a necessary step towards linking platform use to platform materiality. However, this process perspective says little about the role of the human user in shaping outcomes and, consequently, environmental implications. It is therefore essential to continue focusing on how humans shape digital technologies and how this is embedded in a broader socio-environmental framework, emphasising the interplay between human agency, digital platforms and the environment.

The ubiquity and global scale of digital platforms are deeply rooted in human use and the underlying societal norms and values that drive their adoption. Environmental resources are extensively consumed because of persistent user demand. Data centres require vast amounts of energy, while cultural practices add to growing communication demands. For example, the widespread use of messaging services significantly contributes to resource consumption. Multimedia content, such as photos and videos, requires storage in data centres and energy for transmission. Routine practices, such as messaging multiple recipients or duplicating data across several devices, increase the use of environmental resources. While each individual instance of communication might seem minor, the aggregated environmental implication and thus social injustices are substantial. Similarly, e-waste results from both the short lifecycles of devices, typically enforced by design, and human consumption.

Addressing the environmental aspects of digital platforms must therefore move beyond material infrastructure to incorporate the human conditions that shape the development, implementation and use of digital technologies. In this context, research in the field of science and technology studies has demonstrated that technologies are socially constructed and contextually embedded (Bijker & Law, 1994; Lievrouw, 2014). Other studies have shown how developers' and users' narratives, visions and expectations shape the innovation and implementation of digital technologies (Kerr et al., 2020; Llach, 2015). This underscores the importance of considering the human perspective as a shaping force of digital platforms rather than viewing humans as passive consumers. Integrating both platform material and human aspects is a crucial first step towards creating a comprehensive framework that addresses the social, material and environmental dimensions of digital platforms.

Towards Eco-Material Relations

Understanding digital platforms as eco-material relations provides a framework for addressing both the human–technology relation, as conceptualised by Ihde (1990), and the process-oriented understanding of digital platforms' materiality in terms of resources, energy and e-waste, as suggested by Kannengießer and McCurdy (2021). The concept of eco-materiality underscores the interconnectedness of digital platforms with humans and the environment. Focusing on socio-environmental relations moves beyond user-platform dichotomies and human–nature dualisms, arguing for a more integrated view that acknowledges the human–technology–environment entanglements in digital platforms. Beyond that, it emphasises the importance of understanding digital platforms within broader sociocultural frameworks that guide technological development and consumption patterns.

As outlined above, digital platforms can be meaningfully understood through the lenses of alterity and/or background relations. However, in these relations, the 'world', as both a conceptual and an environmental dimension, is viewed as separate and disconnected. To account for this, this chapter introduces a fifth relation – *eco-material relations* (see Fig. 1.2). This relation is characterised by the intentional connection between the human, the platform and the environment. Framing digital platforms as eco-material relations allows the analytical focus to shift towards the way in which users, digital platforms and the environment shape one another. It also enables an investigation of how human agency is exerted within these interconnected processes. Examining digital platforms through the lens of eco-material relations captures both the user-technology relationship and the environmental impacts of these interactions. Thus, digital platforms are conceptualised as a form of eco-material relations in which humans are intentionally linked to the technology, as well as the world, through the elements of resources, energy and e-waste. This means that the use of digital platforms is not only a matter of interaction or communicative relation but also a dynamic process that ties use to the broader environmental implications of digital platforms.

Examining digital platforms through an eco-material lens prompts two primary directions for investigation: the social dimension and the material dimension. Beginning with the latter, investigating the material dimension entails examining the sourcing of natural resources required for platform infrastructure(s). This includes studying the ethical implications of mining practices and addressing the social justice issues that arise from the unequal distribution and use of these resources. Furthermore, addressing the issue of energy consumption within the proposed relational framework highlights the pressing need to tackle the increasing energy demands of technologies such as AI. Finally, recognising e-waste

Embodied relations: (I-technology) -> world (technology is emerged)

Hermeneutic relations: I -> (technology-world) (technology is performative)

Alterity relations: I -> technology (-world) (technology is interactive)

Background relations: I -> technology (-world) (technology is contextual)

Eco-material relations: I - technology - world (technology is material)

Fig. 1.2. Towards Eco-Material Relations. *Source:* Own illustration.

as a critical component of eco-material relations facilitates an investigation into the longevity of digital devices. This encompasses not only the physical waste generated but also the broader consequences of overconsumption and the cultural practices that drive rapid technological turnover. Digital labour, resource extraction and wasting practices often occur in regions of the Global South, where marginalised communities face forced displacement, environmental degradation and health risks. While the economic opportunities associated with these practices may provide short-term benefits, they come at the cost of long-term social and environmental harm. Digital workers in countries such as Venezuela, Brazil, Madagascar and the Philippines perform low-paid, invisible data annotation and moderation tasks that are essential to AI systems (Casilli et al., 2025). What may be seen as profit and convenience for users worldwide often comes at a significant cost to vulnerable communities. Rather than framing socio-environmental impacts as inevitable consequences of societal digitalisation, a critical perspective needs to pay attention to how global inequalities shape the distribution of both environmental impacts and technological benefits.

Continuing with the former, the social dimension involves exploring the human practices, values and societal norms that shape and are shaped by digital platforms. This includes examining how the design of digital platforms influences social norms, such as the normalisation of constant connectivity and the cultural expectation of being 'always on'. Furthermore, it involves exploring the socio-cultural drivers behind platform adoption and use, which are motivated in part by convenience and entertainment, and how these practices contribute to resource consumption and environmental costs. The social dimension also necessitates considering questions of agency and reflectivity in relation to how aware users are of the environmental implications of digital platforms, and where they place responsibility for the impacts (Mahnke et al., 2024).

Concluding Remarks

Digital platforms have become an integrated part of everyday life, and their socio-environmental implications are becoming a critical challenge for digital societies. Addressing these implications should be a priority on current research agendas. As a contribution to digital platform studies, this chapter develops the lens of eco-materiality as a way of reconceptualising digital platforms as eco-material relations. The purpose of exploring such a relational framework is to position human–technology relations within the context of their socio-environmental implications. Central to these relations is the recognition of humans as active agents and key actors. Further, it is essential to recognise that the use of resources for digital technologies disproportionately impacts communities in the Global North and the Global South, which underlines the importance of a justice dimension in studying these implications.

Viewing digital platforms through the lens of eco-materiality demands a shift in how environmental responsibility and accountability are constructed around digital platforms. When humans are understood not merely as consumers of digital platforms (Mahnke et al., 2024) but as active participants in shaping these platforms, as well as their environmental impact, they must become aware of and take responsibility for their actions. While the 'greening' of technological infrastructures in the form of transitioning to renewable energy sources is a necessary step, it must be supported and supplemented by societal awareness and change. Environmental change can only occur together *with* technology. A one-sided focus on either technology or human users hinders the understanding of relational aspects and leads to blind spots. Examining digital platforms through a relational, human-centred approach allows for a broader and more nuanced understanding of the human, technology and environment entanglements.

However, a relational view does not advocate for individualising responsibility. Simply asking users to 'act more responsibly' would overlook the systemic causes of environmental harm, which are deeply embedded in the structural aspects of digital platforms. A shift towards a relational framework is therefore not about reducing human–technology–environmental relations to individual relations; rather, it requires a collective approach that involves rethinking the design of digital platforms, fostering systemic change within the tech industry and promoting a more nuanced understanding of the human–technology–environment nexus.

On this basis, the hope is to empower digital societies to find meaningful solutions to the socio-environmental challenges posed by digital platforms. This includes technological solutions such as circular approaches and design choices that prioritise sustainability and user empowerment, supporting and fostering a culture in which individuals can take responsibility for the environmental impact of their digital practices and allow for green(er) choices as a consequence. At the same time, future research should explore how to address systemic issues, examining not only how to change individual behaviours but also how to influence organisational practices, policy frameworks and societal norms that shape digital platform relations. One example could entail incorporating interface interventions designed to raise awareness of the environmental implications, encouraging individuals to reflect more deeply and critically on their consumption habits. At the same time, such an intervention would provide the tech industry and policymakers with clearer insights into the actual environmental costs of digital platforms. In addition, initiatives aimed at extending the lifecycle of digital devices could support a cultural shift towards greater sustainability in digital practices. Finally, fostering research and educational programmes that raise awareness and inspire action can empower digital societies to better understand these challenges and take meaningful steps to address the environmental costs associated with digital platforms.

References

Batmunkh, A. (2022). Carbon footprint of the most popular social media platforms. *Sustainability*, *14*(4). Article 4. https://doi.org/10.3390/su14042195

Bijker, W. E., & Law, J. (1994). *Shaping technology/building society: Studies in sociotechnical change*. The MIT Press. https://mitpress.mit.edu/9780262521949/shaping-technology-building-society/

Bonini, T., & Treré, E. (2024). *Algorithms of resistance: The everyday fight against platform power*. The MIT Press. https://doi.org/10.7551/mitpress/14329.001.0001

Borning, A., Friedman, B., & Gruen, D. (2018). What pushes back from considering materiality in IT? In *Proceedings of the 2018 workshop on computing within limits* (pp. 1–6). https://doi.org/10.1145/3232617.3232627

Casilli, A. A., Tubaro, P., Cornet, M., Le Ludec, C., Torres-Cierpe, J., & Braz, M. V. (2025). Global inequalities in the production of artificial intelligence. In *The handbook of digital labor* (pp. 217–232). John Wiley & Sons, Ltd. https://doi.org/10.1002/9781119981831.ch12

Coeckelberg, M. (2021). Narrative responsibility and artificial intelligence. *AI & Society*, *38*, 2437–2450. https://doi.org/10.1007/s00146-021-01375-x

Crawford, K. (2021). *Atlas of AI: Power, politics, and the planetary costs of artificial intelligence*. Yale University Press. https://doi.org/10.12987/9780300252392

Cressman, D. (with Cressman, D.). (2022). *The necessity of critique: Andrew Feenberg and the philosophy of technology* (1st ed., Vol. 41). Springer International Publishing AG. https://doi.org/10.1007/978-3-031-07877-4

Dastin, J. (2024, January 16). OpenAI CEO Altman says at Davos future AI depends on energy breakthrough. *Reuters*. https://www.reuters.com/technology/openai-ceo-altman-says-davos-future-ai-depends-energy-breakthrough-2024-01-16/

Ferreboeuf, H., Efoui-Hess, M., & Verne, X. (2021). *Environmental impacts of digital technology: 5-year trends and 5G governance*. The Shift Project. https://theshiftproject.org/wp-content/uploads/2023/04/Environmental-impacts-of-digital-technology-5-year-trends-and-5G-governance_March2021.pdf

Flyverbom, M., Deibert, R., & Matten, D. (2019). The governance of digital technology, big data, and the internet: New roles and responsibilities for business. *Business & Society*, *58*(1), 3–19. https://doi.org/10.1177/0007650317727540

Gabrys, J. (2011). *Digital rubbish*. University of Michigan Press.

Gunkel, D. (2012). Communication and artificial intelligence: Opportunities and challenges for the 21st century. *Communication + 1*, *1*(1), 1–25. https://doi.org/10.7275/R5QJ7F7R

Haider, J., Rödl, M., & Joosse, S. (2022). Algorithmically embodied emissions: The environmental harm of everyday life information in digital culture. *IR Information Research*, *27*(special issue). https://doi.org/10.47989/colis2224

Hao, K. (2019). *Training a single AI model can emit as much carbon as five cars in their lifetimes*. MIT Technology Review. https://www.technologyreview.com/2019/06/06/239031/training-a-single-ai-model-can-emit-as-much-carbon-as-five-cars-in-their-lifetimes/

Helberger, N., Pierson, J., & Poell, T. (2018). Governing online platforms: From contested to cooperative responsibility. *The Information Society*, *34*(1), 1–14. https://doi.org/10.1080/01972243.2017.1391913

Ihde, D. (1990). *Technology and the lifeworld*. Indiana University Press. https://iupress.org/9780253205605/technology-and-the-lifeworld/

Jones, N. (2018). How to stop data centres from gobbling up the world's electricity. *Nature*, *561*(7722), 163–166. https://doi.org/10.1038/d41586-018-06610-y

Kannengießer, S. (2018). Repair cafés as communicative figurations: Consumer-critical media practices for cultural transformation. In A. Hepp, A. Breiter, & U. Hasebrink (Eds.), *Communicative figurations: Transforming communications in times of deep mediatization* (pp. 101–122). Springer International Publishing. https://doi.org/10.1007/978-3-319-65584-0_5

Kannengießer, S. (2020). Engaging with and reflecting on the materiality of digital media technologies: Repair and fair production. *New Media & Society*, *22*(1), 123–139. https://doi.org/10.1177/1461444819858081

Kannengießer, S., & McCurdy, P. (2021). Mediatization and the absence of the environment. *Communication Theory*, *31*(4), 911–931. https://doi.org/10.1093/ct/qtaa009

Kerr, A., Barry, M., & Kelleher, J. D. (2020). Expectations of artificial intelligence and the performativity of ethics: Implications for communication governance. *Big Data & Society*, *7*(1). https://doi.org/10.1177/2053951720915939

Kimball, S. (2024, May 15). *Microsoft's carbon emissions have risen 30% since 2020 due to data center expansion*. CNBC. https://www.cnbc.com/2024/05/15/microsofts-carbon-emissions-have-risen-30percent-since-2020-due-to-data-center-expansion.html

Kneese, T., & Young, M. (2024). Carbon emissions in the tailpipe of generative AI. *Harvard Data Science Review*, (special issue, 5). https://doi.org/10.1162/99608f92.fbdf6128

Kuntsman, A., & Rattle, I. (2019). Towards a paradigmatic shift in sustainability studies: A systematic review of peer reviewed literature and future agenda setting to consider environmental (un)sustainability of digital communication. *Environmental Communication*, *13*(5), 567–581. https://doi.org/10.1080/17524032.2019.1596144

Lai, S. S., & Flensburg, S. (2023). *Gateways: Comparing digital communication systems in Nordic welfare states*. Nordicom, University of Gothenburg. https://doi.org/10.48335/9789188855848

Lievrouw, L. A. (2014). Materiality and media in communication and technology studies: An unfinished project. In T. Gillespie, P. J. Boczkowski, & K. A. Foot (Eds.), *Media technologies: Essays on communication, materiality, and society* (pp. 21–51). The MIT Press. https://doi.org/10.7551/mitpress/9780262525374.003.0002

Llach, D. C. (2015). *Builders of the vision: Software and the imagination of design*. Routledge. https://doi.org/10.4324/9781315798240

López, A., Ivakhiv, A., Rust, S., Tola, M., Chang, A. Y., & Chu, K. (Eds.). (2024). *The Routledge handbook of ecomedia studies*. Routledge, Taylor & Francis Group. https://www.routledge.com/The-Routledge-Handbook-of-Ecomedia-Studies/Lopez-Ivakhiv-Rust-Tola-Chang-Chu/p/book/9781032009421

Lucivero, F. (2019). Big data, big waste? A reflection on the environmental sustainability of big data initiatives. *Science and Engineering Ethics*, *26*(2), 1009–1030. https://doi.org/10.1007/s11948-019-00171-7

Mahnke, M. S., & Bagger, C. (2024). Navigating platformized generative AI: Examining early adopters' experiences through the lens of data reflectivity. *Convergence*, *30*(6), 1974–1991. https://doi.org/10.1177/13548565241300857

Mahnke, M. S., Pruulmann-Vegerfeld, P., Swart, J., & Mathieu, D. (2024). Data reflectivity and user reflexivity: New conceptual pathways in bridging data/platform and user/audience studies. *Convergence*, *30*(6), 1859–1870. https://doi.org/10.1177/13548565241301459

Malmodin, J., Lövehagen, N., Bergmark, P., & Lundén, D. (2024). ICT sector electricity consumption and greenhouse gas emissions – 2020 outcome. *Telecommunications Policy*, *48*(3), 102701. https://doi.org/10.1016/j.telpol.2023.102701

Mathieu, D., & Pruulmann-Vengerfeldt, P. (2020). The data loop of media and audience: How audiences and media actors make datafication work. *MedieKultur: Journal of Media and Communication Research*, *36*(69). https://doi.org/10.7146/mediekultur.v36i69.121178

Maxwell, R. (2014). Media industries and the ecological crisis. *Media Industries Journal*, *1*(2). https://doi.org/10.3998/mij.15031809.0001.207

Natarajan, U. (2021). Environmental justice in the Global South. In C. G. Gonzalez, S. L. Seck, & S. A. Atapattu (Eds.), *The Cambridge handbook of environmental justice and sustainable development* (pp. 39–57). Cambridge University Press. https://doi.org/10.1017/9781108555791.005

Pickard, V. (2021). A new social contract for platforms. In M. Moore & D. Tambini (Eds.), *Regulating Big Tech: Policy responses to digital dominance* (pp. 323–337). Oxford University Press. https://doi.org/10.1093/oso/9780197616093.003.0018

Plantin, J.-C., Lagoze, C., Edwards, P. N., & Sandvig, C. (2016). Infrastructure studies meet platform studies in the age of Google and Facebook. *New Media & Society*, *20*(1). https://doi.org/10.1177/1461444816661553

Raghavan, B., & Ma, J. (2011). The energy and emergy of the internet. In *Proceedings of the 10th ACM workshop on hot topics in networks* (pp. 1–6). https://doi.org/10.1145/2070562.2070571

Rice, J. (2007). Ecological unequal exchange: Consumption, equity, and unsustainable structural relationships within the global economy. *International Journal of Comparative Sociology*, *48*(1), 43–72. https://doi.org/10.1177/0020715207072159

Roos, A., Kostakis, V., & Giotitsas, C. (2016). Introduction: The materiality of the immaterial: ICTs and the digital commons. *tripleC: Communication, Capitalism & Critique. Open Access Journal for a Global Sustainable Information Society*, *14*(1), 48–50. https://doi.org/10.31269/triplec.v14i1.738

Rosignoli, F., Privitera, E., & Pappalardo, G. (2023). Earth system at risk: Challenging environmental (in)justice. *Environmental Justice*, *16*(3), 171–173. https://doi.org/10.1089/env.2023.29016.frs

Saenko, K. (2023, May 23). Is generative AI bad for the environment? A computer scientist explains the carbon footprint of ChatGPT and its cousins. *The Conversation*. http://theconversation.com/is-generative-ai-bad-for-the-environment-a-computer-scientist-explains-the-carbon-footprint-of-chatgpt-and-its-cousins-204096

Schäfer, M. S., & Schlichting, I. (2014). Media representations of climate change: A meta-analysis of the research field. *Environmental Communication*, *8*(2), 142–160. https://doi.org/10.1080/17524032.2014.914050

Schwartz, S. A., & Mahnke, M. S. (2021). Facebook use as a communicative relation: Exploring the relation between Facebook users and the algorithmic news feed. *Information, Communication & Society*, *24*(7), 1041–1056. https://doi.org/10.1080/1369118X.2020.1718179

Sharma, P., & Dash, B. (2022). The digital carbon footprint: Threat to an environmentally sustainable future. *International Journal of Computer Science and Information Technology*, *14*(03), 19–29. https://doi.org/10.5121/ijcsit.2022.14302

Stephansen, H., & Treré, E. (2020). *Citizen media and practice: Currents, connections, challenges*. Routledge: Taylor & Francis Group. https://www.routledge.com/Citizen-Media-and-Practice-Currents-Connections-Challenges/Stephansen-Trere/p/book/9781138571846

van Dijck, J. (2013). *The culture of connectivity: A critical history of social media* (1 ed.). Oxford University Press.

Verbeek, P. P. C. C. (2001). Don Ihde: The technological lifeworld. In H. J. Achterhuis (Ed.), *American philosophy of technology: The empirical turn* (pp. 119–146). Indiana University Press. https://research.utwente.nl/en/publications/don-ihde-the-technological-lifeworld

Vestberg, R. M., Raundalen, J., & Lager, N. (Eds.). (2014). *Media and the ecological crisis*. Routledge. https://doi.org/10.4324/9781315885650

Wang, H., Yang, G., & Yue, Z. (2023). Breaking through ingrained beliefs: Revisiting the impact of the digital economy on carbon emissions. *Humanities and Social Sciences Communications*, *10*(1), 1–13. https://doi.org/10.1057/s41599-023-02126-7

Chapter 2

The Challenging Task of Estimating the Carbon Footprint of Digital Services

Salla-Maaria Laaksonen[a] *and Gopika Premsankar*[b]

[a]University of Helsinki, Finland
[b]Aalto University, Finland

Abstract

While hardware becomes more energy-efficient, the growing use of digital services results in increasing energy consumption for the information and communications technology (ICT) sector. This increase also raises concerns about the carbon footprint of producing, hosting and consuming digital services. Furthermore, digital services such as social media platforms and streaming services rely on compute- and data-intensive content formats, ranging from high-resolution media to the increasing use of generative artificial intelligence, which further contribute to the growth of carbon emissions from such services. In this chapter, the authors describe the main components that cause carbon emissions when providing digital services, namely, end-user devices, communication networks and data centres. Next, the authors analyse the carbon emissions reported by six major platforms that provide such services. To conclude, the authors highlight the main challenges in estimating the carbon emissions of digital services due to the complexity of the ecosystem, lack of information and lack of enforced standards and accountability obligations. The transition towards more sustainable digital platforms needs to involve developers, engineers, service providers, hosting platforms as well as consumers.

Keywords: Carbon emissions; carbon footprint; sustainability reporting; platform companies; cloud infrastructure; digital services

Platforms and the Planet, 37–51

doi:10.1108/978-1-83662-172-020261003

Introduction

> [W]e will have to monetize it somehow at some point; the compute costs are eye-watering. (OpenAI CEO Sam Altman on Twitter/X, December 5 2022)

The hidden environmental impacts of digitalisation have started to be highlighted by researchers, non-governmental organisations and the media in recent years. Digital services have been perceived as intangible, which hides their energy- and material-intensive nature. Recent studies have highlighted the indirect impact of end-user devices, servers and digital infrastructure on the biosphere and climate (Andrae, 2020; Istrate et al., 2024; Nardi et al., 2018; Obringer et al., 2021; Olsson et al., 2023; Preist et al., 2016; Røpke, 2012). Current evidence suggests that the use of information and communications technology (ICT) is growing faster than its energy efficiency (Belkhir & Elmeligi, 2018; Freitag et al., 2021; Goldman Sachs, 2024). This development is further intensified with the use of artificial intelligence (AI) and large language models, which require increasing computation and data centre capacities (Kamiya & Coroamă, 2025; Kneese & Young, 2024). On a broader scale, the environmental impact of digital platforms is also a question of global inequality and climate justice, as the concrete consequences of growing use result in extreme weather conditions, natural resource extractivism and appropriation of land and resources particularly in the Global South and formerly colonised countries (e.g., Andreucci et al., 2023; Arora, 2016). Therefore, solutions are needed to address the environmental sustainability of digital technologies. In the pursuit of such a goal, the field needs reliable information about the environmental consequences related to the use of digital services, in particular their carbon footprint. Such information is also needed because of regulations that require companies to report carbon emissions from the use of their products and services (European Commission, n.d.).

Measuring carbon emissions from digital services is difficult because of the complexity of measuring electricity consumption across the entire production and consumption chain, and the lack of a widely adopted approach that would also cover the downstream emissions resulting from service use. For example, estimation of the emissions from viewing a video on a smartphone requires data from digital service providers, platform companies and infrastructure providers, as well as localised data on electricity consumption, telecommunications network design, consumer terminals and connections. Therefore, adopting the widely used Greenhouse Gas (GHG) Protocol for carbon accounting for digital services has methodological uncertainties that can lead to variations in carbon reporting across organisations (Gritsenko et al., 2024; Microsoft, 2023). There are, nevertheless, some user-facing services for estimating the carbon footprint resulting from an individual's or organisation's use of digital and cloud services (AWS, 2024; Google, 2024a; Greenspector, 2023; Microsoft, 2024a; see also Vigren et al., *this volume*). While such tools provide some insight into the environmental impact from the use of digital services, they are dependent on the user behaviour and choices, such as number of queries made, amount of data stored, screen resolution used and so on. These measurements have to be added as assumptions to the estimation models, which essentially introduces uncertainties in the estimations provided. We argue that it is essential to also quantify the carbon emissions from the overall use of a digital service, collectively across all users. Such estimates help to identify the digital services that contribute most to carbon emissions and, consequently, the ones that must be focused on first to reduce overall emissions from the ICT sector.

In this chapter, we provide an overview of the existing approaches to estimate the carbon emissions resulting from digital services and discuss the difficulties in the process. We focus on the operational carbon emissions, that is, the carbon emissions associated with the electricity consumed for accessing and running digital services (Bashir et al., 2023). We first discuss the global status and development trends of the carbon emissions of digital services. After that, we review the components that need to be taken into account when estimating the energy consumption and associated operational emissions. Next, we explain the reporting scheme used for GHG emissions and how it has been applied in the reporting of six major digital platform companies: three cloud service companies, i.e., Google, Amazon and Microsoft, and three digital content service providers, i.e., Meta, Spotify and Netflix. We conclude by discussing the challenges related to measuring and estimating emissions for individual digital services. To conclude, we propose some ways forward to achieve more robust measurements and to mitigate the climate impact of digital services.

Growing Energy Consumption of Digital Services

A major source of carbon emissions from digital services originates from the energy they consume. At the time of writing, we lack a total and separate account of the energy consumption of services provided by the digital platform industry, but we know that the entire ICT industry uses 7–9% of the world's electricity consumption (Andrae, 2020; World Bank & International Telecommunications Union, 2024). The resulting share of carbon emissions from the ICT sector in global emissions is estimated to be between 1.5% and 4% (World Bank & International Telecommunications Union, 2024), which exceeds the share of the frequently discussed aviation sector at 2% (International Energy Agency, 2023a). While the energy efficiency of consumer devices and data centres has developed thanks to regulation (e.g. the Energy-Related Products Directive of the European Union (EU) and Energy Star in the USA; Freitag et al., 2021), there are two counterforces to this development – the growing use of digital services and increasingly energy-intensive forms of content.

In 2024, we spent on average over six hours a day globally with digital devices (DataReportal, 2024). The number of Internet users doubled between 2014 and 2024 (Statista, 2024). In 2022, video content accounted for 65% of all Internet consumer traffic (Sandvine, 2023). Videos, audio, multimedia, immersive and interactive content formats require more energy to produce, store and transfer than text content (e.g., Morley et al., 2018). An analysis of the 1,000 most popular Finnish websites showed that their size had multiplied during the preceding 10 years, in particular for the sites intended for mobile devices because of higher-resolution images and videos, interactive content and the increased use of web analytics (Saarinen & Manner, 2022).

Further, the growing use of AI and large language models also requires more computing resources and energy (e.g., Crawford, 2024; Dhar, 2020; Goldman Sachs, 2024; Luccioni et al., 2024). There seems to be a general institutional tendency in digital infrastructure design to embed novel technologies and features, such as generative AI tools, in customer products at a fast pace, which also poses requirements for the computing infrastructure, network bandwidth and, consequently, energy use (Olsson et al., 2023; Olsson, *this volume*; Preist et al., 2016; Varoquaux et al., 2024). Both Google and Microsoft announced in 2024 that their GHG emissions were on the rise and communicated uncertainty about whether they will reach their climate goals because of AI training and deployment in their data centres (Milmo, 2024). AI is expected to be the key driver for the growing energy use in data centres in the near future (Kamiya & Coroamă, 2025), and technology giants are already planning new fossil fuel infrastructures to prepare for this rapid development (Elliott, 2025; Wong, 2025). This will have further consequences on water and land use as well, and it will further incentivise big tech companies to expand their power in the management of natural resources (Hogan, 2018).

In light of these trends, it becomes clear that, like other consumption, digital content consumption should also be in the spotlight when striving for a more sustainable and just future. In the next section, we will unpack the components of digital services in order to understand what causes carbon emissions when using the services.

Components of Digital Services' Emissions

The operational emissions of a digital service are estimated based on the energy consumed by all the components of the service, which are then multiplied by the carbon intensity of the energy supply used to power those components. However, estimating emissions of digital services is a difficult task because of the complexity of the value chain. The sources of carbon emissions for digital services can be broadly classified into three components: consumer devices (or end-user devices) used to utilise the digital services, the networks over which the services are accessed (i.e. mobile and fixed networks) and cloud data centres, where the services are typically run (World Bank & International Telecommunications Union, 2024). Fig. 2.1 shows an overview of the components.

Consumer devices

This component comprises the carbon emissions associated with electricity consumed by devices such as smartphones, tablets, laptops and desktops while

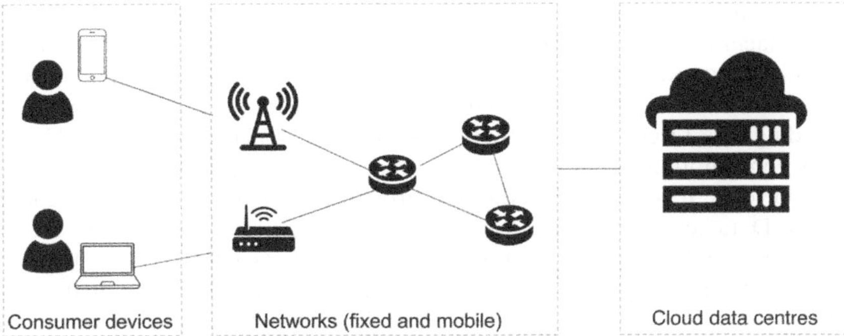

Fig. 2.1. Sources of Emissions for Digital Services.

using digital services. The processing capabilities of these devices and screen size are primary factors in determining the energy consumed. Some reports show that the energy consumption and carbon emissions of consumer devices such as televisions are the biggest contributors to overall emissions when consuming digital content (DIMPACT, 2022; Netflix, 2024).

Networks

End users access the digital services over a communication network, comprising mainly either mobile or fixed networks. Such networks comprise infrastructure such as routers, base stations and content delivery network servers to deliver digital content and services to end users. Although the type of content (text, image, video) affects how much data is used to download the content, the energy consumption of the network infrastructure is not directly proportional to the volume of traffic. Moreover, network data transmission technologies have a big impact on energy consumption and, therefore, also on carbon emissions. Fixed networks are currently a more energy-efficient choice than mobile, and newer mobile network technologies are more efficient than older ones (Huttunen et al., 2023). The electricity consumption of communication networks, both fixed and wireless, was estimated to account for about 1–1.15% of global electricity use in 2022 (International Energy Agency, 2023b). Nearly two-thirds of this consumption comes from mobile communication networks (International Energy Agency, 2023b), and mobile data traffic is expected to almost triple by 2030 (Ericsson, 2024).

Data centres

The global data centre infrastructure – comprising facilities with compute servers, storage servers and networking equipment – supports the running of digital services (Mytton & Ashtine, 2022). Such facilities are a key component of the carbon emissions associated with digital services as they consume a large amount of energy to keep the IT equipment running in addition to the energy required for the facilities (for lighting, cooling the servers, etc.). The growing energy demand of data centres is one of the most significant drivers of carbon emissions in the ICT sector globally (International Energy Agency, 2024; Milmo, 2024). According to a recent estimate by the International Energy Agency (2024), the energy consumption of data centres, cryptocurrencies and AI will double in the next two years. It is unclear whether this growth can be met by renewable energy sources. Data centre operators typically claim to operate on renewable energy, but often greening is achieved through carbon offsets (Brevini, 2022). In the wake of AI, many big tech companies have already pushed their climate goals further into the future (Milmo, 2024) and have expressed interest in nuclear power (Penn & Weise, 2024) as well as fossil fuel-based sources (Elliott, 2025).

The estimation of carbon emissions for digital services is thus complicated by the fact that there are several components in the entire end-to-end pipeline. In the next section, we discuss how carbon emissions are reported by the platform companies involved in providing digital services.

Standards and Current Status of Reporting

ICT companies typically report the development of their carbon footprint annually in accordance with the GHG Protocol, an international standard for reporting carbon emissions (World Business Council for Sustainable Development & World Resources Institute, 2004). According to the standard, it is customary to classify emissions into Scopes 1, 2 and 3, which are further divided into subcategories. The classification of emissions helps organisations understand where their own emissions originate. In particular, companies calculate and report direct emissions from their own operations (Scope 1), emissions from the production of purchased energy (Scope 2) and partly also indirect emissions from the value chain (Scope 3). In the operation of digital technology companies, the vast majority, typically more than 90%, are indirect Scope 3 emissions from the value chain (World Bank & International Telecommunications Union, 2024). These include, for example, purchased products and services, and the use of products manufactured by the company. When evaluating the environmental effects of digital technology, the key measurable object is Category 11 of Scope 3 emissions – the use of sold products. Currently, the ICT industry seems to lack consensus on what this actually covers in digital services. For example, the GSM Association et al. (2023) guidance for measuring Scope 3 emissions instructs telecommunication companies to measure end-users' device electricity consumption but skips emissions arising from sold services as they 'are more complex to assess' (p. 60).

This aspect makes emissions assessment and related activities pronouncedly complicated for digital services – the indirect emissions are the most significant, but measuring them is particularly challenging, lacking commonly accepted measurement methods and a shared understanding of who is responsible for measuring the different components. This applies to some extent to all fields but perhaps most notably for digital services (Vieira et al., 2024). At the time of writing, the EU has introduced the Corporate Sustainability Reporting Directive (CSRD) (European Commission, n.d.), which forces large companies operating in the EU to report their Scope 3 emissions. The schedule is now a bit more complicated thanks to the new Omnibus package. This forces most digital services providers, such as cloud provider platforms and digital content platforms, to find ways to estimate their indirect emissions.

Reporting Practices by Platform Companies

The platform companies release their environmental reports detailing the carbon emissions associated with their operations, but their reporting practices vary. All six companies we explored – Google, Amazon, Microsoft, Meta, Netflix and Spotify – show similar patterns in their emission structures. The majority originate from Scope 3 emissions, ranging from 75% to 96.5%. The share of Scope 1 emissions is minimal – around 1% for Google (2024b) and Microsoft (2024b), but 21% for Amazon (2024), which reports the emissions associated with all business activities jointly, including delivery and logistics related to e-commerce. For digital content providers, the numbers are even higher – Scope 3 emissions account for 98.3% of all emissions by the music streaming company Spotify (2024) and 99% of the emissions by Meta (2024), a provider of several digital services with a global user base.

Major cloud providers Google and Microsoft report growth in Scope 2 emissions from their data centres in 2024. According to the GHG Protocol, Scope 2 emissions can be location or market based. Location-based emissions comprise emissions associated with the electricity sources at each location of the data centre, regardless of the renewable energy agreements used by cloud providers (Maji et al., 2024). On the other hand, market-based emissions account for renewable energy purchases made by the cloud provider. This means that cloud providers can report reduced carbon emissions by investing in renewable energy even if their data centres are not actually powered by green energy (Maji et al., 2024). This option is extensively used by the ICT sector; for example, Amazon only reports market-based Scope 2 emissions (Amazon, 2024).

Google's (2024b) report includes the location-based and market-based emissions by their data centres but does not provide a breakdown of emissions by provided consumer-facing digital services (e.g., Google Search, Google Drive) or for AI (e.g., Gemini). Amazon (2024) reports the market-based Scope 2 emissions associated with electricity purchased for Amazon's operations but does not report them separately for their data centres. Thus, a key challenge remains in identifying the emissions associated with specific digital services, such as Amazon Web Services (AWS).

Google, Microsoft and Amazon provide tools to measure the carbon footprint of using their cloud services, targeted towards B2B users (AWS, 2024; Google, 2024a; Microsoft, 2024a). The tools report carbon emissions according to guidelines defined in the GHG Protocol – while Google and Microsoft tools report all three scopes, AWS's Customer Carbon Footprint Tool only reports emissions related to Scopes 1 and 2. The numbers reported are then used by digital services providers who use cloud services from Google, Microsoft or Amazon.

Thus, reporting practices are complicated and obscured by the intertwined nature of the platform infrastructure. Spotify and Netflix provide streaming services without operating their own data centres (Netflix, 2024; Spotify, 2024). Rather, they purchase the relevant compute and storage services from large cloud providers such as Google (Google, 2024c) and AWS (Netflix, 2024). Netflix reports both their market- and location-based emissions from all three scopes (Netflix, 2024). Their Scope 3 emissions include the use of sold products, but the use of cloud services is reported only as market based because of the reporting structures inherited from AWS. Consequently, Netflix (2024) reports only 5% of its carbon footprint for the streaming process ('streaming delivery infrastructure and data centre providers'). An exceptional aspect of Netflix is that the main share of their overall carbon emissions stems from production activities – the physical activities behind digital content production.

Challenges and Uncertainties in the Estimation

Estimating the energy consumption and related emissions from digital services is a complex task, and common reporting schemes are yet to be made. As shown in the previous section, a major chunk of the emissions results from Scope 3 emissions, which introduces uncertainty and complexity into the modelling – the high impact of product use emissions requires information also from other components of the system. Furthermore, as several actors are involved in the value chain, companies will have to rely on reporting provided by others. In such a setting, it becomes difficult to pinpoint who is and should be accountable for the emissions and to define how societies should mitigate them. This is obvious when scrutinising the various practices platform companies use to report their emissions. Companies publish information as part of their own environmental reporting, if at all. For example, there is no public information available regarding the electricity consumption or emissions of the AI giant OpenAI.

Koomey and Masanet (2021) discuss common pitfalls faced when aiming to analyse or assess the energy consumption of digital services. They point out that any model made is inevitably a simplification of the real-world situation. Models typically rely on retrospective data and therefore fail to account for rapid changes in the system, in particular related to energy-efficiency improvements. Consequently, when estimates aim for long-term projections, they tend to overestimate energy use and underestimate technological change and efficiency improvements. Further, the authors remind us that most models rely on measuring data traffic. However, changes in data traffic are not linearly reflected in energy use because there are also other influencing factors (see also World Bank and International Telecommunications Union, 2024). Measurement and reporting, however, have to rely either on activity-based data or data from other partners in the supply chain. When reports are required, companies have to turn to whatever data are available.

These pitfalls are present with the existing models and tools developed with an aim to estimate the carbon emissions of digital services taking into account all components (see Fig. 1). Such tools suggest ways to estimate the impact of a consumer's (see Vigren et al., *this volume*) or media producer's choices on total energy consumption and carbon emissions. They are typically analytical models or simulations that have to make assumptions about the choices of different components during the consumption process. For example, Greenspector (2023), which measures the carbon footprint of social media services, assumes the user's location, consumer device, screen brightness and server location in its calculations. Another model is the DIMPACT (2022) tool, widely used by media companies, which takes into account three levels of measurement for data transmission – servers (service provider and third party), edge and backbone networks and the access network used by the customer – and, as a fourth level, the consumer device used by the customer (see also Schien et al., 2013). Again, all these measurements need to be adapted to the local and national use environment. However, both technology and its usage patterns are constantly developing, which is why the calculations become quickly outdated. Nevertheless, they might provide useful heuristics for understanding the structure of emissions resulting from digital services.

Despite the inherent uncertainties, the estimates might have discursive power. As Koomey and Masanet (2021) remind us, overestimations or miscalculations of energy use and the climate impact of ICT easily become 'factoids' – compelling stories that gain media attention. For example, in 2019, a claim circulated in the media stating that an hour of video streaming was equivalent to driving a car for 4 miles (6.4 kilometres) in terms of carbon emissions (for background and correction, see The Shift Project, 2020). In Finland, a similar story compared streaming a movie with heating an electric sauna in terms of energy consumption (Mattila, 2022), which was then later heavily disputed. As methodologies for estimating carbon emissions resulting from digital services are improving, such stories cast unnecessary doubt on the entire endeavour. As catchy headlines, however, they are effective in generating awareness.

Conclusion

In this chapter, we have discussed the need for and challenges of estimating the climate impact of digital services offered by platform companies. As the consumption of digital services grows rapidly, we should pay attention to low-carbon ways of consuming digital services. While we know that estimating the climate effects of digital technologies is imbued with uncertainties related to technological change, sociotechnical complexities and the volatile nature of digitalisation processes (Gritsenko et al., 2024), estimates will be made in the coming years because of regulatory requirements. A key requirement is to have generally accepted and sufficiently accurate ways of estimating the carbon emissions of digital services, which, in turn, would help in raising awareness and in designing regulations. Building such measurement standards is a collaborative project that requires the committed cooperation of the entire digital platform industry and stakeholders, including regulators.

Most notably, there is a need for more transparent sharing of data by the industry to gain more accurate data on which to base the estimates. Much of the necessary data and statistics are behind technology companies (Crawford, 2024; Pasek et al., 2023). There are some initiatives and discussions about fostering more sustainable digital infrastructures by the companies (see Frig et al., *this volume*; Hogan et al., *this volume*). The digital infrastructure climate coalition iMasons Climate Accord, for example, is an initiative by major tech companies to advocate for open governance to report the carbon impact of materials, products and power across the infrastructure of data centres. At the same time, it is clear that a focus on the materiality of data centres is not enough, but the discussions should also cover the usage of the digital end products and, for that, estimating emissions is a necessity. In this regard, the CSRD requirement to report Scope 3 emissions is a good start, which forces companies to look beyond their own premises. More accurate information on the environmental impact of digital services can help environmental agencies and decision-makers to monitor the use trends on an aggregate level, which provides better accounts for regulating the field. And, perhaps, the reports can result in consumer awareness if the companies are willing to implement design solutions that would make this information visible to users of digital services. We are already used to seeing the CO_2 estimates when buying flights, but could there be a similar number available when browsing Netflix, for example?

However, we stress that the transition to more sustainable digital platforms cannot be a consumer responsibility only, but a structural issue (see also Vigren et al., *this volume*) that requires action from the digital platform industry and regulators. Numbers and quantification tend to be essential in igniting action in the context of business and government (Porter, 1995), which further highlights the need for legitimate estimates of the environmental impact of digital services. Yet, acknowledging the complexity of developing the emission estimates, we echo the thoughts of Gritsenko et al. (2024) and advocate for broadening the focus into discussions concerning solutions and reduction mechanisms as well. Gritsenko et al. (2024) argue that because of the inherently unpredictable measurement of digital carbon footprints, societies should focus on mitigation rather than mere prediction of these effects. Their proposed mitigation strategies include transition to on-site green energy production and better management of digitalisation processes. Others have suggested identifying less carbon-intensive elements of digital networks and moving the focus to developing those (Pasek et al., 2023).

Correspondingly, we wish to highlight the need to strive for solutions to address the sustainability issues related to digital services. We cannot wait for accurate estimates to be achieved before moving forward with action. To generate such action, the process needs to involve developers, engineers, service providers, hosting platforms, as well as consumers. Therefore, we present a call for further research to understand the complex technical, infrastructural, organisational and behavioural questions related to the sustainability transition of digital services. Such knowledge is essential to identify where regulatory or voluntary interventions would be most effective.

Acknowledgements

This work has been supported by The Media Industry Research Foundation of Finland and Kone Foundation.

References

Amazon. (2024). *2023 Amazon sustainability report*. https://sustainability.aboutamazon.com/2023-amazon-sustainability-report.pdf

Andrae, A. (2020). New perspectives on internet electricity use in 2030. *Engineering and Applied Science Letters*, *3*(2), 19–31.

Andreucci, D., López, G. G., Radhuber, I. M., Conde, M., Voskoboynik, D. M., Farrugia, J. D., & Zografos, C. (2023). The coloniality of green extractivism: Unearthing decarbonisation by dispossession through the case of nickel. *Political Geography*, *107*, 102997. https://doi.org/10.1016/j.polgeo.2023.102997

Arora, P. (2016). Bottom of the data pyramid: Big Data and the global south. *International Journal of Communication*, *10*(0), 19. https://ijoc.org/index.php/ijoc/article/view/4297

AWS (Amazon Web Services). (2024). *Customer carbon footprint tool*. https://aws.amazon.com/aws-cost-management/aws-customer-carbon-footprint-tool/

Bashir, N., Irwin, D., Shenoy, P., & Souza, A. (2023). Sustainable computing — Without the hot air. *ACM SIGENERGY Energy Informatics Review*, *3*(3), 47–52.

Belkhir, L., & Elmeligi, A. (2018). Assessing ICT global emissions footprint: Trends to 2040 & recommendations. *Journal of Cleaner Production*, *177*, 448–463. https://doi.org/10.1016/j.jclepro.2017.12.239

Brevini, B. (2022). *Is AI good for the planet?*. John Wiley & Sons.

Crawford, K. (2024). Generative AI's environmental costs are soaring — and mostly secret. *Nature*, *626*, 693. https://doi.org/10.1038/D41586-024-00478-X

DataReportal. (2024). *Digital 2024: Global overview report*. https://datareportal.com/reports/digital-2024-global-overview-report

Dhar, P. (2020). The carbon impact of artificial intelligence. *Nature Machine Intelligence*, *2*(8), 423–425. https://doi.org/10.1038/s42256-020-0219-9

DIMPACT. (2022). *Methodology: Estimating the carbon impacts of serving digital media and entertainment products*. https://dimpact.org/resource?resource=2

Elliott, R. F. (2025). Chevron joins race to generate power for A.I. *New York Times*. https://www.nytimes.com/2025/01/28/business/energy-environment/chevron-power-plant-ai.html

Ericsson. (2024). *Ericsson mobility report November 2024*. https://www.ericsson.com/en/reports-and-papers/mobility-report/reports/november-2024

European Commission. (n.d.). *Corporate sustainability reporting*. https://finance.ec.europa.eu/capital-markets-union-and-financial-markets/company-reporting-and-auditing/company-reporting/corporate-sustainability-reporting_en

Freitag, C., Berners-Lee, M., Widdicks, K., Knowles, B., Blair, G. S., & Friday, A. (2021). The real climate and transformative impact of ICT: A critique of estimates, trends, and regulations. *Patterns*, *2*(9). https://doi.org/10.1016/j.patter.2021.100340

Frig, M., Dovbysh, O., & Laaksonen, S.-M. (this volume). *Greening platforms or organisational bullshit? Typology of empty and misleading environmental communication by global platform companies.*

Google. (2024a). *Carbon footprint reporting methodology*. https://cloud.google.com/carbon-footprint/docs/methodology

Google. (2024b). *2024 environmental report*. https://www.gstatic.com/gumdrop/sustainability/google-2024-environmental-report.pdf

Google. (2024c). *Google cloud featured customer: Spotify*. https://cloud.google.com/customers/featured/spotify

Greenspector. (2023). *Social carbon footprint calculator*. https://greenspector.com/en/what-is-the-environmental-footprint-of-social-networking-applications-2023/

Gritsenko, D., Aaen, J., & Flyvbjerg, B. (2024). Rethinking digitalization and climate: Don't predict, mitigate. *Npj Climate Action*, *3*, 43. https://doi.org/10.1038/s44168-024-00127-z

GSM Association, the Global Enabling Sustainability Initiative and the International Telecommunication Union. (2023). *Scope 3 guidance for telecommunication operators*. https://www.gsma.com/solutions-and-impact/connectivity-for-good/external-affairs/gsma_resources/scope-3-guidance/

Hogan, M. (2018). Big data ecologies. *Ephemera*, *18*(3), 631–657.

Hogan, M., Verhoeven, D., & Brown, T. J. (this volume). *The archival paradox: DNA-based data storage and the data hoarding crisis.*

Huttunen, J., Pärssinen, M., Heikkilä, T., Salmela, O., Manner, J., & Pongracz, E. (2023). Base station energy use in dense urban and suburban areas. *IEEE Access*, *11*, 2863–2874. https://doi.org/10.1109/ACCESS.2023.3234192

International Energy Agency. (2023a). *Aviation*. https://www.iea.org/energy-system/transport/aviation

International Energy Agency. (2023b). *Data centres and data transmission networks*. https://www.iea.org/energy-system/buildings/data-centres-and-data-transmission-networks

International Energy Agency. (2024). *Electricity 2024. Analysis and forecast to 2026*. Revised version. https://www.iea.org/reports/electricity-2024

Istrate, R., Tulus, V., Grass, R. N., Vanbever, L., Stark, W. J., & Guillén-Gosálbez, G. (2024). The environmental sustainability of digital content consumption. *Nature Communications*, *15*(1), 3724. https://doi.org/10.1038/s41467-024-47621-w

Kamiya, G., & Coroamă, V. C. (2025). Data centre energy use: Critical review of models and results. In *EDNA – IEA 4E TCP*. International Energy Agency (IEA). https://www.iea-4e.org/wp-content/uploads/2025/05/Data-Centre-Energy-Use-Critical-Review-of-Models-and-Results.pdf

Kneese, T., & Young, M. (2024). Carbon emissions in the tailpipe of generative AI. *Harvard Data Science Review, Special Issue*, *5*. https://doi.org/10.1162/99608f92.fbdf6128

Koomey, J., & Masanet, E. (2021). Does not compute: Avoiding pitfalls assessing the internet's energy and carbon impacts. *Joule*, *5*(7), 1625–1628. https://doi.org/10.1016/j.joule.2021.05.007

Luccioni, S., Jernite, Y., & Strubell, E. (2024). Power hungry processing: Watts driving the cost of AI deployment?. In *FAccT '24: Proceedings of the 2024 ACM conference on fairness, accountability, and transparency* (pp. 85–99). https://doi.org/10.1145/3630106.3658542

Maji, D., Bashir, N., Irwin, D., Shenoy, P., & Sitaraman, R. K. (2024). The green mirage: Impact of location-and market-based carbon intensity estimation on carbon optimization efficacy. In *e-Energy '24: Proceedings of the 15th ACM international conference on future and sustainable energy systems* (pp. 256–267). https://doi.org/10.1145/3632775.3639587

Mattila, I. (2022, December 28). Elokuvan katsominen mobiili-liittymän kautta vastaa jopa saunan lämmitystä, vaikka se ei näy sähkölaskussa [watching a movie via your mobile connection is equivalent even to heating a sauna, but it doesn't show up on your electricity bill]. *Helsingin Sanomat*. https://www.hs.fi/kulttuuri/art-2000009293361.html

Meta. (2024). *2024 Sustainability report*. https://sustainability.atmeta.com/wp-content/uploads/2024/08/Meta-2024-Sustainability-Report.pdf

Microsoft. (2023). *Carbon accounting at Microsoft*. https://query.prod.cms.rt.microsoft.com/cms/api/am/binary/RW13XCo

Microsoft. (2024a). *Emissions impact dashboard*. https://www.microsoft.com/en-us/sustainability/emissions-impact-dashboard

Microsoft. (2024b). *2024 Environmental sustainability report*. https://query.prod.cms.rt.microsoft.com/cms/api/am/binary/RW1lMjE

Milmo, D. (2024). Google's emissions climb nearly 50% in five years due to AI energy demand. *The Guardian*. https://www.theguardian.com/technology/article/2024/jul/02/google-ai-emissions

Morley, J., Widdicks, K., & Hazas, M. (2018). Digitalisation, energy and data demand: The impact of internet traffic on overall and peak electricity consumption. *Energy Research & Social Science*, *38*, 128–137. https://doi.org/10.1016/J.ERSS.2018.01.018

Mytton, D., & Ashtine, M. (2022). Sources of data center energy estimates: A comprehensive review. *Joule*, *6*, 2032–2056.

Nardi, B., Tomlinson, B., Patterson, D. J., Chen, J., Pargman, D., Raghavan, B., & Penzenstadler, B. (2018). Computing within limits. *Communications of the ACM*, *61*(10), 86–93. https://doi.org/10.1145/3183582

Netflix. (2024). *Environment social governance report 2023*. https://ir.netflix.net/governance/ESG/default.aspx

Obringer, R., Rachunok, B., Maia-Silva, D., Arbabzadeh, M., Nateghi, R., & Madani, K. (2021). The overlooked environmental footprint of increasing internet use. *Resources, Conservation and Recycling*, *167*, 105389. https://doi.org/10.1016/j.resconrec.2020.105389

Olsson, T., Pyyhtinen, O., Laaksonen, S.-M., Rantasila, A., Vigren, M., Ylipulli, J., & Sawhney, N. (2023). A user-centered lens into digital excess: Exploring the superfluity and environmental burden of the digital world. In *Proceedings of the LIMIT ninth computing within limits 2023*. https://doi.org/10.21428/bf6fb269.a5916a92

Pasek, A., Vaughan, H., & Starosielski, N. (2023). The world wide web of carbon: Toward a relational footprinting of information and communications technology's climate impacts. *Big Data and Society*, *10*(1). https://doi.org/10.1177/20539517231158994

Penn, I., & Weise, K. (2024). Amazon, Google and Microsoft are investing in nuclear power. *New York Times*. https://www.nytimes.com/2024/10/16/business/energy-environment/amazon-google-microsoft-nuclear-energy.html

Porter, T. M. (1995). *Trust in numbers: The pursuit of objectivity in science and public life*. Princeton University Press. https://doi.org/10.1017/CBO9781107415324.004

Preist, C., Schien, D., & Blevis, E. (2016). Understanding and mitigating the effects of device and cloud service design decisions on the environmental footprint of digital infrastructure. In *CHI '16: Proceedings of the 2016 CHI conference on human factors in computing systems* (pp. 1324–1337). https://doi.org/10.1145/2858036.2858378

Røpke, I. (2012). The unsustainable directionality of innovation – The example of the broadband transition. *Research Policy*, *41*(9), 1631–1642. https://doi.org/10.1016/J.RESPOL.2012.04.002

Saarinen, A., & Manner, J. (2022). *Analyysi suosituista Suomessa käytetyistä verkkosivuista* [An analysis of popular websites used in Finland]. Aalto-yliopiston *julkaisusarja [Aalto University publication series] TIEDE + TEKNOLOGIA 1/2022*. http://urn.fi/URN

Sachs, G. (2024). *AI is poised to drive 160% increase in data center power demand*. https://www.goldmansachs.com/insights/articles/AI-poised-to-drive-160-increase-in-power-demand

Sandvine. (2023). *Global internet phenomena report*. https://www.sandvine.com/press-releases/sandvines-2023-global-internet-phenomena-report-shows-24-jump-in-video-traffic-with-netflix-volume-overtaking-youtube

Schien, D., Shabajee, P., Wood, S. G., & Preist, C. (2013). A model for green design of online news media services. In *Proceedings of the 22nd international conference on world wide web* (pp. 1111–1122). https://doi.org/10.1145/2488388.2488485

Spotify. (2024). *Equity and impact report 2023*. https://www.lifeatspotify.com/reports/Spotify-Equity-Impact-Report-2023.pdf

Statista. (2024). *Number of internet users worldwide from 2005 to 2024*. https://www.statista.com/statistics/273018/number-of-internet-users-worldwide/

The Shift Project. (2020). *Did the shift project really overestimate the carbon footprint of online video? Our analysis of the IEA and CarbonBrief articles*. https://theshiftproject.org/en/article/shift-project-really-overestimate-carbon-footprint-video-analysis

Varoquaux, G., Luccioni, A. S., & Whittaker, M. (2024). *Hype, sustainability, and the price of the bigger-is-better paradigm in AI*. https://arxiv.org/abs/2409.14160v1

Vieira, L. C., Longo, M., & Mura, M. (2024). Impact pathways: The hidden challenges of Scope 3 emissions measurement and management. *International Journal of Operations & Production Management*, *44*(13), 326–334. https://doi.org/10.1108/IJOPM-01-2024-0049

Vigren, M., Karppi, T., & Pyyhtinen, O. (this volume). *Calculating, reducing, compensating and disconnecting: Summons of digital excess*.

Wong, M. (2025). The false AI energy crisis. *The Atlantic*. https://www.theatlantic.com/technology/archive/2025/02/ai-energy-crisis-fossil-fuels/681653/

World Bank and International Telecommunications Union. (2024). *Measuring the emissions & energy footprint of the ICT sector: Implications for climate action*. https://www.itu.int/hub/publication/d-ind-clim-2023-01/

World Business Council for Sustainable Development and World Resources Institute. (2004). *The greenhouse gas protocol: A corporate accounting and reporting standard.* Revised Edition. https://ghgprotocol.org/sites/default/files/standards/ghg-protocol-revised.pdf

Chapter 3

Understanding Moral Lapses in Design: Cultural Biases, Beliefs and Narratives as Factors of Accidental Evil

Thomas Olsson

Tampere University, Finland

Abstract

This chapter discusses the moral responsibility of the designers of online platforms, mindful of the diverse effects of information technology on individuals, society and the environment. Designers are trained experts in problem-solving who outline visions of sociotechnical futures and shape innovative digital artefacts. In other words, they exert social and political power, which ought to come with responsibility. This chapter bridges literature across disciplines to refocus attention on the cultural drivers of design work. It analyses cultural biases, beliefs and narratives as factors of *accidental evil*, i.e., moral lapses where unexpected repercussions materialise despite the designers' benign intentions. Three factors – *techno-solutionism*, *grand narratives* and *collective myopia* – are discussed in the context of online platforms and theoretically unpacked in relation to design responsibility and design justice. The analysis offers considerations for the development of competencies supporting critical thinking and prefigurative risk assessment to avoid unexpected harm to society and the planet.

Keywords: Design responsibility; design justice; design praxis; IT ethics; sustainable digitalisation; accidental evil; online platforms; techno-solutionism; grand narratives; risk analysis

Platforms and the Planet, 53–66

doi:10.1108/978-1-83662-172-020261004

Introduction

The multifarious systemic effects of digitalisation and information technology (IT) on individuals, society and the environment have been long discussed in, for example, science and technology studies, IT ethics and media studies. As outlined by the introduction of this volume, a recent target of moral concern are the adverse effects of digital media platforms and on-demand video streaming. Despite the increased convenience digital platforms offer, we ought to reconsider not only extensively problematised issues like disinformation (Westlund, 2021) or information addiction (Lozano-Blasco, 2022) but also the growing material costs resulting from intensive computation and data transfer (Istrate et al., 2024).

In fact, the 2010s and 2020s could be regarded as a consolidation of several classical truisms of technological change and its effects on human affairs. For example, in his 1998 talk, Neil Postman argued that the benefits and costs of technology are never distributed evenly among the population, and that technological change is neither *additive* nor *incremental* but *ecological* and *systemic*. Melvin Kranzberg (1986) underlined that technologies are morally *non-neutral*: while the morality of a technological solution partly depends on its uses, its capability to afford certain malicious uses also demands attention (for example, consider generative artificial intelligence in the present). It has become ever clearer that online platforms and their highly optimised algorithms have also changed the *background* in which people act together (Winograd & Flores, 1986).

The increasing volume of critical literature on this topic and the breadth of the materialised harms have engendered debate and self-reflection in technology professionals' communities, particularly those related to human–computer interaction and interaction design. An increasing body of design literature considers the power, effects and underlying values of design, such as in design philosophies and movements like Critical Design (Dunne & Raby, 2001), Posthumanist Design (Forlano, 2017) and Design Justice (Costanza-Chock, 2020). One reflection of this development is the birth of interdisciplinary and technology-critical conferences like ACM FAccT and Computing within Limits, and collectives like the Design Justice Network that aim to challenge structural inequalities enabled and encouraged by design. In practitioner communities, professional ethics guidelines seem to be increasingly revisited by societies like the Association for Computing Machinery (ACM) and the Interaction Design Association (IxDA). I regard all these developments as an ongoing *ethical awakening*, that is, a collective process of recognising the professional community's moral positions and responsibility vis-a-vis society and the environment.

This chapter aims to accelerate this momentum of professional ethics discussion and to support the recalibration of the moral compass in design work. To understand the roles of design as a profession, I argue for the importance of recognising and critically assessing certain cultural mindsets and values deeply embedded in the culture of design and technology development in general. The focus is on detrimental cultural influences that may deviate from individual designers' benign intentions and moral standards – i.e., *moral lapses* where one act against their better judgement. In other words, the discussed influences may result in so-called *accidental evil.* According to Nelson and Stolterman (2012), accidental evil is both *unintended* and *unexpected* – harm created by misfortune, accident or the fact that the designer's power is employed without understanding the possible environmental or societal repercussions. This differs from *wilful evil* as *intentional* use of power in design for malign purposes, such as misconduct by using dark design patterns to nudge certain user behaviours. I analyse three examples of cultural effects as enablers of accidental evil: (1) the diverse forms of techno-solutionism, (2) grand narratives and beliefs in the IT sector and (3) collective myopia regarding the evolution of complex systems.

Let me clarify my positionality and methodological approach to help interpret the weight and relevance of the arguments in this chapter. First, it is noteworthy that the analysis is based on a combination of accumulated professional experience as a design educator, collaboration with design professionals in the IT sector and the reflections of various literature across disciplines. The presented factors are to be read as discussion openers or tentative hypotheses rather than the results of a systematic analysis of empirical evidence. Second, being educated in computer science and human–computer interaction and having taught designers for more than 15 years, I am both a member and product of the broad designer community that this article addresses. Third, my personal and professional values reflect those of Nordic welfare states. In design, IT and engineering, these values reflect long traditions of participatory design and open-source development, and a culture built on social democracy, pragmatism, a flat power hierarchy and a growing sustainability mindset. Against this backdrop, I aim not to reprimand design professionals but to provoke self-reflection and engender discussion within the professional community that I deeply care for and know the best. This community has only existed for a few decades and therefore lacks a long history of professional identity clarification.

In what follows, I first outline a general background and philosophical fundamentals of design, after which the three cultural factors of accidental evil are thoroughly discussed one by one. The chapter concludes with remarks on how to reconsider and mitigate the cultural factors in design practice.

Power, Responsibility and Evil in Design

> The first is the ubiquity of design – design is literally everywhere; from the largest structures to the humblest aspects of everyday life, modern lives are thoroughly designed lives. (Escobar, 2017, p. 2)

All design is concerned with change and preferred futures, be it about the design of physical products, digital artefacts or online platforms. Design utilises and produces knowledge for future transformations, defining alternative future-oriented directions to solve real-world problems. At a more practical level, Nelson and Stolterman (2012, p. 6) summarise that 'design is the ability to imagine that-which-does-not-yet-exist, to make it appear in concrete form as a new purposeful addition to the real world.' These gradually lead to a realisation of the inherently ethico-political essence of design. Mindful of this particularly in industrial design, already Victor Papanek (1972/1991, p. ix) provocatively argued that 'there are professions more harmful than industrial design, but only a very few of them.'

My focus is on design professionals who shape digital services, applications and artefacts – experts with professional titles like 'service designer', 'user experience designer' or 'product owner'. This relatively young and broad professional community plays key roles in inventing digital applications and matching new technical capabilities with human desires. They craft service concepts, customer journeys and user interfaces, typically aiming at stellar user experiences to maximise customer satisfaction and user engagement. Especially those in leading positions are central to the *raison d'etre* of technological applications: they define the visions, goals and key features, thus setting conditions for people's experiences and behaviour for years to come.

Such expertise indeed implies a position of power that designers and design organisations ought to acknowledge. As all artefacts may embody political power (Winner, 1980), for example, through the affordance to benefit or discriminate against certain groups, the design of digital artefacts for specific futures and aims is also a fundamentally moral and political action. Building on the tradition of Value-Sensitive Design, Davis et al. (2015) highlight how technological influence is a sum of multiple factors, including the features of a technological design and the people responsible for them. Importantly, such indirect power introduces concerns. While an invention might appear justifiable due to its benign intentions and expected benefits, all solutions may bring about unpredictable and indirect negative outcomes in the long term. Escobar (2017) regards design as *worldmaking*, which is inherently political action, and calls for design practice that is attuned to global justice and environmental reflection.

Recognising this power to influence people and have indirect effects on sociocultural and environmental systems has engendered discussions on designers' moral responsibility to humankind. For example, Mike Monteiro (2019), a long-term design practitioner in this area, suggested that the profession needs hard moral codes: while lawyers and doctors can lose their licenses upon failing to follow ethical standards, designers might get a bonus for their daring and innovative trials. Don Norman (2024), a founding father of usability engineering and long-term advocate of user-centred design, has called for redeveloping design praxis towards an idea where success is measured not in money but in the wellness and happiness of all people. These problematisations and realisations of moral responsibility within the community are further indications of the ethical awakening mentioned earlier.

The idea that design should be responsive to society and the environment is not a recent concern, as underlined by, for example, socially responsible design (Melles et al., 2011). The authors point to Papanek's classic book *Design for the Real World* (1972/1991), which crystallised many of the societal and environmental concerns into an argument for change through approaches like eco-design and inclusive design. However, considering the short history of designing digital artefacts, it is understandable that we are now seeing a renaissance of discourses around responsibility, justice and sustainability.

The following sections outline three cultural factors of accidental evil that I regard as strongly intertwined and particularly important for explaining why unintended consequences emerge through the design of digital services. My focus on *cultural aspects* contrasts with various *cognitive* pitfalls discussed in prior literature. For example, Selbst et al. (2019) identify five traps in which one might fall when attempting to innovate digital solutions for complex social problems. For example, the *formalism trap* is about the failure to account for the full meaning of social concepts, as they can be contextual and contestable and out of reach of mathematical formalisms. The following factors partly relate to cognitive aspects, too, but their importance as factors of accidental evil is emphasised through the *cultural, intersubjective aspects.*

Techno-Solutionism: A Biased Attitude Towards the Potency of Technology and Designers

Techno-solutionism refers to a way of thinking that all problems can be addressed by technology – a 'confidence in the human capacity to find and implement numerous strategies for overcoming obstacles and capitalizing on opportunities' (Sætra & Selinger, 2023, p. 9). Conceptually, it builds on the general idea of solutionism, a term used intermittently since the mid-1900s, and recently re-introduced into the public discourse (e.g., Morozov, 2013). Solutionism presents a bias not only among technologists but also in policymaking and jurisdiction: for example, considering policies and regulations as prime solutions to injustice in societies. This reminds us that the approaches and instruments an individual knows best often dictates their general approach to problem-solving (Selbst et al., 2019). Techno-solutionism is also closely related to the more historical notion of *techno-fixes*, which implies a belief that simple problems may be resolved easily and quickly with technological patches (Sætra & Selinger, 2023).

In the context of designing digital applications, I regard techno-solutionism through three facets. First, it is driven by a collective bias driving consideration of which wrongs can and ought to be remedied. For example, the development of digital platforms for romantic partner searches implies a technocratic idea that complex social problems can be fixed by simply introducing a digital marketplace.

Second, it reflects a learnt attitude towards design experts' capabilities to address all kinds of issues. For example, IT professionals may also regard social issues as open for discovery, just like natural phenomena. This reflects a strongly positivist idea that may further bolster the belief in engineers' and designers' omnipotent capabilities to address any complex issue. Such excessive confidence may lead to courageous and somewhat techno-imperialist attempts to digitalise anything, anywhere.

Third, it is built on hopeful beliefs that all difficulties have benign solutions, and that computation is a potent solution approach with minimal negative impact. For example, the present-day hype over artificial intelligence reflects very techno-optimist and enthusiastic attitudes (Danaher, 2022), wherein we have the potential to engineer new societies with the wonders of new technology and radically renew old conventions in all areas of life. A related belief is the *de-materiality* of digital solutions, underlining the lack of negative effects; the design of digital services might appear free from material conditions and environmental repercussions, but it is now apparent that this is hardly the truth (Crawford, 2024; Olsson et al., 2023; Mahnke, *this volume*; Bartlett & Velkova, *this volume*).

While the digitalisation attempts fuelled by techno-solutionism may be pursued for genuinely benevolent and universalist intentions, the flip side is that new harms generated from design interventions tend to be overlooked. Returning to the nature of technological change, the produced solutions tend to result in ripple effects and new problems, some of which may be more harmful than the original problem. A key issue appears to be that such truisms about the nature of technological change are often overshadowed by the professional confidence and enthusiasm that also drive techno-solutionism.

Considering design as a professional praxis, I regard techno-solutionism to result from – and contribute to – several aspects of present-day design work. For example, a solution-driven mindset is in fact an inherent element in the definitions of design we covered earlier: design is fundamentally regarded as problem-solving. In other words, this can be seen as a fundamental human bias prominent among designers; everything is considered *designable* and to have a *telos*, an ultimate purpose. Design is both necessary for innovation and risky when applied to any given problem, but the former aspect tends to be emphasised: problem-solving is pursued at the cost of analysis of the contextual nuances and cultural undercurrents. Moreover, when coupled with the productional expectations towards design work, priority tends to be given to intuitive solutions and quick fixes. In contrast, many design philosophers (e.g., Nelson & Stolterman, 2012) and interdisciplinary scholars (e.g., Becker, 2023; Costanza-Chock, 2020) imply the importance of problem definition, analysis and goal setting in all design work. Therefore, to avoid the follies of techno-solutionism, designers could first ask if the perceived problems are the most relevant problems worth addressing and do they afford verifiable solutions. Building on Baumer and Silberman (2011), it can be difficult to recognise that the best solution – for society or the planet – might not involve any digital elements.

Another connection to modern design work is the insufficiency of merely technical solutions to address complex sociocultural, organisational and political issues. Morozov (2013) problematises techno-solutionism as a tendency to recast complex social situations as neat processes that can be easily optimised with computational solutions. Here, the notion of wicked problems (Alford & Head, 2017) is highly relevant, referring to essentially unique problems that cannot be exhaustively formulated and that have no immediate test of a solution (e.g., climate change). The authors argue that 'we do not so much "solve" wicked problems as make progress towards improvement or towards better managing them' (Alford & Head, 2017, p. 1). Moreover, wicked problems are socially constructed and highly contextual. Mitra et al. (2024) argue that the underlying uncritical techno-optimistic narrative neglects such complexities and nuances of wicked problems. Techno-solutionist tendencies therefore hinder the analysis of the nature and premises of a given problem statement.

Finally, techno-solutionism seems to be reinforced by the externalisation of responsibility. Designers tend to externalise the origin of problems to other professions (for example, policymakers or bureaucrats; Morozov, 2013). Simultaneously, other professions tend to outsource problem-solving to designers: for many domain specialists, it is appealing to outsource problem-solving by simply investing money in a design firm and thus be able to present a modern, technologically sophisticated intervention. This arrangement of professional specialisation separates those who solve problems from those who do not. Among designers, this can lead to a sense of professional superiority over others, which further reinforces the belief of being able to solve any issue. For example, the very idea of 'hero designers', omnipotent experts whose intuition leads to the best possible outcomes, may imply symptoms of professional hubris.

Underlying Grand Narratives and Beliefs in the Professional Culture

The next broad factor is the grand narratives and values in the IT sector, influencing a broad community including not only designers but also many other professionals. In general, cultural narratives are the stories and beliefs that groups of people lean on to establish a professional identity, support a sense of community, construct meaning in their work and define moral boundaries (Saleebey, 1994). In IT, a plethora of narratives at different scales and spanning different periods can be identified: from long-term narratives about technological progress to more recent narratives about impartial algorithms free from human judgement biases. However, here the focus is on the most profound stories, ideas and values that have shaped the societal and economic status of the IT sector. Such grand narratives are built into the very fundaments of the professional practice, manifesting as common beliefs, ideals, expectations, skillsets and methods, and an overall ethos – they are the cultural cornerstones of the global design community.

Perhaps the most dominant narrative in design relates to its position in society. If science aims to describe *the world as it is*, design aims to describe *how the world could be* (Nelson & Stolterman, 2012). In the collective mentality, the field is entitled to the power to invent and give form to the future and define people's ways of living on the planet within the industrial system (Celaschi et al., 2019). This could partly explain why many designers and engineers are more interested in getting things to work than considering the reasoning behind the technologies they use. Coupled with the techno-solutionist attitude, this collective mentality has concretised into catchphrases like *Move fast and break things; unless you are breaking stuff, you are not moving fast enough.* This reflects the ideals of agility, radical innovations, a trial-and-error mindset and technological progress that describe the ethos of the 2000–2010s. The same intention to fundamentally transform traditional industries is captured by the phrase *Software is eating the world* (attributed to software businessperson Marc Andreessen), demonstrating a growth mindset and an idolisation of the culture of technology-driven innovation to that end. The cultural effects of these attitudes appear central to the present-day values in design work, and there are hardly any competing narratives that would be equally widespread.

Following Christoph Becker's (2023) thinking, such narratives can be regarded as examples of the common myth of technocracy and technological determinism present in this field. Technological determinism espouses the idea that technical progress follows a unilinear course, free from influence from the society that it shapes (Feenberg, 2010). It regards technical progress as an *exogenous force* shaping society, rather than an expression of changes in culture and values, which contrasts with view that cultural ideas and values shape both technologies and the people creatively appropriating them (Feenberg, 2010). The power to induce change through technology that is built into determinism may indeed seem appealing to people whose main instrument is IT.

Dominant narratives tend to develop into an underlying belief system, and these beliefs may range from epistemic questions such as the nature of truth to elements of mythology and religion, for example, fictions about what will remain after we pass. Such deeper philosophical ideas have recently been discussed in IT about movements such as *transhumanism*, *longtermism* and *extropianism*. Transhumanistic aspirations include utilising advanced technologies to enhance human nature and continue our biological evolution towards the posthuman (Akpan, 2024). Building on futurist Ray Kurzweil's science fiction and transhumanistic ideas, present-day examples of programmes that reflect this ideology include artificial superintelligence and digital solutions to augment human senses and capabilities. Extropianism can be considered a version of transhumanism that emphasises the continuous improvement of the human condition with the help of science, ultimately letting people live indefinitely. Further, longtermism is founded on the idea that positively influencing the long-term future (thousands to millions of years) is a moral priority (MacAskill, 2022). This idea can be recognised in, for example, the ambitions of humans becoming a multi-planetary species. In terms of religious connotations, such philosophies and movements tend to lean on universally desired ideas such as evading human extinction, reaching enlightenment, settling new worlds or even religious thoughts about resurrection and immortality.

But why is this problematic? While the goals may appear sensible to many, leaning on such fundamental human desires and fears imposes the risk of them gradually becoming justifications for any technology-development ambitions. In IT and design, these could include endeavours that are disproportionally expensive or environmentally harmful in the short term, for example. In other words, a key risk is that influential people may use such philosophies as smoke and mirrors to deceive masses of highly capable technology developers, possibly disguising their true intentions. It is noteworthy that the specific *intended* ends of bold endeavours often remain unspecified. Likewise, the means to reach the aims often appear equally vague. For example, in the pursuit of artificial general intelligence, there is no clarity or consensus on either the aims or the means.

Placing too much faith in vague long-term benefits tends to reflect a narrow view of moral consideration, namely the *consequentialist logic* where benign ends justify any means. Blindly following this logic ignores other relevant schools of thought in moral philosophy, such as following certain universal moral duties and imperatives (*deontology*) or virtue ethics, which focuses on the goodness of the character of key actors, such as designers.

Now, how do the abovementioned narratives and beliefs manifest in the present-day design of online platforms? One such form is the general primacy of technological novelty over any other value. Another is the focus on convenience and ease in user experience design. While these values appear universally desirable in the tradition of human-centred design, optimising for them can undermine the consideration of the moral grounds and long-term effects. Moreover, such values tend to be concretised into organisational metrics and performance indicators, which solidifies and institutionalises the values, eventually leading to professional standards that are difficult to question.

Another key effect is the pace of design and development work, fuelled by the intrinsic value assigned to agility and dynamism. For example, to avoid overspending on a low-value idea, it is key to *fail fast*, and to showcase the solution idea as early as possible one may *fake it till you make it* – both ideas actively employed particularly by IT startups. A knock-on effect of such values is the tendency to follow general design trends, mimic ideas from successful products and resort to design patterns and conventions. It is no surprise that digital media platforms have resulted in a massive increase in energy consumption: mimicry in design practice tends to produce more of what there already is. For example, digital media platforms appear to uncritically copy features from each other, including video reels and ephemeral stories, which allows for the reproduction and exacerbation of the harms of past designs.

A related tendency is incremental development. Following the agile development manifesto in the mid-2000s, a sprint-based and iterative approach to designing and developing software has become an industry norm. This means that, even if repeated revisioning of the product idea appears sound in principle, it can be practically challenging to make significant changes of direction during the development process. For example, the sunk cost fallacy and groupthink effects may set economic and social barriers to escape a certain trajectory or product idea. While incremental development typically works well in the context of information systems with clear instrumental aims and limited uses, applying this practice to complex and open-ended sociotechnical systems such as online platforms demands much more thought-through solution proposals to start with. Moreover, the common use of quick and dirty methods, for example, heuristic analysis and quick group reflections, typically called *retrospects*, may undermine ethical deliberation. The lack of thoroughness in the methods undermines the capability to prefigure long-term systemic effects.

All in all, technical interventions built on the narratives and values described above introduce significant new risks that often remain unrecognised, largely because of the narratives themselves. Innovation-led, disruption-seeking narratives can also include pathological elements when the values and long-term aims remain unclear. For example, individuals, organisations and practices might need to drastically adjust to the new technological conditions, retrospectively fixing the solution and making it appear successful. The examples above call for re-establishing the dominant narratives and beliefs in the IT sector. Alternative mindsets could be applied from, for example, safety-critical domains where the toleration for error is minimal.

Collective Myopia Regarding the Evolution of Complex Systems

The third factor is how our understanding is inherently limited with respect to how the sociotechnical systems that the artefacts we design are part of will evolve in the long term. This fundamentally cognitive limitation, short-sightedness or *myopia* in figurative terms, is present in all human activities, shaping our understanding of the long-term ramifications and complex chains of causalities into the future. Importantly, the capability to predict future actions is not only challenged by our cognitive limits or the mere complexity of sociotechnical systems but also by the cultural narratives discussed above. Therefore, we also need to recognise a cultural phenomenon where short-sightedness is fuelled by techno-solutionist mindsets, lack of methods for prefiguration and the cultural beliefs and narratives discussed above. This cultural layer could be termed *collective myopia*.

In design and development work, collective myopia may be present in cases of creative appropriation of earlier ideas and when building on prior solutions. The previous section established that designers may uncritically follow trends and mimic prior innovations. Uncritical mimicry introduces a portability trap (Selbst et al., 2019), where one might fail to understand how repurposing a solution designed for a specific context may be harmful, inappropriate or inaccurate when applied to another context. On the other hand, the adoption of prior solutions may also be very creative, and existing systems may be adopted for various purposes. This refers to gradual function expansion (or function creep), where the system is imperceptibly expanded beyond its original purposes, causing contestable change (Koops, 2021). A recent example would be the scale-up of large language models (LLMs) beyond specific instrumental uses, such as translation or predictive text entry, to more computationally intensive and ubiquitous forms of text generation. This may also include contestable uses such as deception, political influencing or the fabrication of textual evidence, also contributing to the growing environmental footprint of LLMs. In other words, *qualitative changes* in a system's function may cause *quantitative and harmful changes* in the outcomes, and recognising such causalities in advance is very challenging.

Moreover, the creative appropriation of digital technology is also pursued by the users (Kim & Lee, 2012), where people creatively utilise a platform for their instrumental, emotional, social and curiosity-driven needs. The massive scale of, for example, a social online platform significantly expands the possible uses beyond what the designers originally considered. Predicting all the diverse and emergent uses is, understandably, extremely difficult. As a result, the related environmental costs and societal harms may not be pre-emptively addressed. Rather, they may simply emerge unexpectedly and can only be retroactively mitigated, often with insufficient means (for example, reactive content moderation on social platforms). Moreover, digital services afford cyclic chain reactions: certain user behaviours influence the service, which affords new user behaviours, further shaping the service. Such emergent loops may amplify both positive and negative effects in unpredictable ways.

Another aspect of collective myopia is how design is organised and what methods are employed for prefiguration and risk management. Organising design work is typically driven by the search for efficiency and the management of systemic complexity. Individual designers or teams tend to work on only parts of the whole, which can lead to so-called *local optimising* of specific use cases or key performance indicators. A concrete example is the designers' lack of visibility to the material costs: they might not have good visibility to the full technical stack (especially the Scope 3 emissions from the use of cloud services, see Premsankar & Laaksonen in this volume). Therefore, the indirect environmental footprint of specific design choices remains unknown. This emphasises the importance of roles such as product owners and service portfolios to create a collective understanding of the sum of the parts.

It appears self-evident that we need better collective capability to predict and retrospectively manage the evolution of the sociotechnical systems in which our designs exist. Design methodology and literature in human–computer interaction, fortunately, provide some approaches, methods and tools for just this: classical approaches such as scenario work and participatory design through diverse teams, as well as cross-disciplinary practices for anticipation through future studies (Hautala & Ahlqvist, 2022) and critical and speculative design (Pierce, 2021). Unfortunately, a recent extensive review of design projects by Schwarz et al. (2023) presents scant evidence on the use of such foresight methods. While the reviewed design projects, from mostly academic contexts, may deviate from day-to-day work by design practitioners, I share the authors' doubts about the systematic use of diverse methods for prefigurative risk analysis. Even if relevant methods and design tools were actively used, another pitfall is an overreliance on them. While moral deliberation and self-reflection can indeed benefit from canvas tools and design checklists, such thinking may also be limited to the viewpoints offered by a given tool, especially if broader reflections are not encouraged.

Concluding Remarks and Implications

Through a cross-disciplinary synthesis of the applicable research literature, this chapter sheds light on three intertwined cultural factors that contribute to the emergence of accidental evil in digital services: techno-solutionism, grand narratives in IT and collective myopia. I consider the discussed factors relevant because they reflect very common yet generally overlooked professional tendencies in design work and more broadly in IT.

The factors also appear inherently intertwined. For example, certain narratives may support techno-solutionism, which may reinforce myopia, which can then undermine our capability to challenge design conventions. Together, they may encourage designers to enthusiastically adopt technological interventions, pursuing technological progress that introduces the risk of innovations that are decoupled from a deeper purpose and lack sustainable value. That said, this is by no means an exhaustive list of cultural factors. For example, the analysis intentionally excludes the influences of prevalent economic models, curricula in computer science education and specific political philosophies prevalent in IT, such as libertarianism as a foundation of the Internet.

Indeed, I believe there is room to improve both in terms of individual practices and the organisational culture and leadership of design work. For example, to fight techno-solutionism, everyone involved in design ought to acknowledge their power and responsibility. We need to nurture a culture of professional humility and more thoughtful processes – a new design praxis that truly prioritises the diversity of humankind and the environment (e.g., Becker, 2023; Costanza-Chock, 2020; Escobar, 2017; Norman, 2024). To compensate for myopia, we need to pursue pluralistic viewpoints, more deliberative practices and new means for perspective taking, anticipation and design critique. To question the dominant narratives and the incumbent values in IT, the design community could develop new narratives that encourage more sustainable digitalisation endeavours and build on ideas like *digital sufficiency* (Santarius et al., 2023) and *design justice* (Costanza-Chock, 2020), rather than maximising user engagement and fuelling the attention economy. To these ends, I recommend incorporating examples of and reflections on accidental evil into design education and human–computer interaction curricula.

I hope this account offers considerations and preliminary directions for the long-term development of competencies in pre-emptive risk assessment and morally sustainable innovation work. Unearthing these cultural factors as pitfalls can help designers avoid moral lapses and act according to a more thought-through value basis. To effectively serve as cautionary advice, I encourage consideration of how these factors may manifest in different applications, projects, organisations, geographic regions and professional contexts.

References

Akpan, T. M. (2024). *Transhumanist technologies as enhancers of human nature and its dignity*. AI Ethics. https://doi.org/10.1007/s43681-024-00559-7

Alford, J., & Head, B. W. (2017). Wicked and less wicked problems: A typology and a contingency framework. *Policy and Society*, *36*(3), 397–413. https://doi.org/10.1080/14494035.2017.1361634

Baumer, E. P. S., & Silberman, M. S. (2011). When the implication is not to design (technology). In *Proceedings of CHI '11* (pp. 2271–2274). ACM. https://doi.org/10.1145/1978942.1979275

Becker, C. (2023). *Insolvent: How to reorient computing for just sustainability*. The MIT Press. http://mitpress.mit.edu/9780262545600

Celaschi, F., Formia, E., Iñiguez Flores, R., & León Morán, R. (2019). Design processes and anticipation. In R. Poli (Ed.), *Handbook of anticipation* (pp. 773–793). Springer. https://doi.org/10.1007/978-3-319-91554-8_48

Costanza-Chock, S. (2020). *Design justice: Community-led practices to build the worlds we need*. MIT Press. https://doi.org/10.7551/mitpress/12255.001.0001

Crawford, K. (2021). *Atlas of AI: Power, politics, and the planetary costs of artificial intelligence*. Yale University Press.

Danaher, J. (2022). Techno-optimism: An analysis, an evaluation and a modest defence. *Philosophy & Technology*, *35*(2), 1–29.

Davis, J., Nathan, L. P., & van den Hoven, J. (2015). Value sensitive design: Applications, adaptations, and critiques. In P. Vermaas & I. van de Poel (Eds.), *Handbook of ethics, values, and technological design* (pp. 11–40). Springer. https://doi.org/10.1007/978-94-007-6970-0_3

Dunne, A., & Raby, F. (2001). *Design noir: The secret life of electronic objects*. Birkhäuser.

Escobar, A. (2017). *Designs for the pluriverse: Radical interdependence, autonomy, and the making of worlds*. Duke University Press. http://www.jstor.org/stable/j.ctv11smgs6

Feenberg, A. (2010). *Between reason and experience: Essays in technology and modernity*. MIT Press. https://doi.org/10.7551/mitpress/8221.001.0001

Forlano, L. (2017). Posthumanism and design. *She Ji: The Journal of Design, Economics, and Innovation*, *3*(1), 16–29. https://doi.org/10.1016/j.sheji.2017.08.001

Hautala, J., & Ahlqvist, T. (2022). Integrating futures imaginaries, expectations and anticipatory practices: Practitioners of artificial intelligence between now and future. *Technology Analysis & Strategic Management*, *36*(9), 2100–2112. https://doi.org/10.1080/09537325.2022.2130041

Istrate, R., Tulus, V., Grass, R. N., Vanbever, L., Stark, W. J., & Guillén-Gosálbez, G. (2024). The environmental sustainability of digital content consumption. *Nature Communications*, *15*(1), 1–11. https://doi.org/10.1038/s41467-024-47621-w

Kim, H., & Lee, W. (2012). Framing creative uses for describing cases of appropriation. In *Proceedings of CSCW '12* (pp. 135–138). ACM. https://doi.org/10.1145/2141512.2141561

Koops, B. J. (2021). The concept of function creep. *Law, Innovation and Technology*, *13*(1), 29–56. https://doi.org/10.1080/17579961.2021.1898299

Kranzberg, M. (1986). Technology and history: "Kranzberg's laws." *Technology and Culture*, *27*(3), 544–560. https://doi.org/10.2307/3105385

Lozano-Blasco, R., Quilez Robres, A., & Soto Sánchez, A. (2022). Internet addiction in young adults: A meta-analysis and systematic review. *Computers in Human Behavior*, *130*. https://doi.org/10.1016/j.chb.2022.107201

MacAskill, W. (2022). *What we owe the future*. Basic Books.

Melles, G., de Vere, I., & Misic, V. (2011). Socially responsible design: Thinking beyond the triple bottom line to socially responsive and sustainable product design. *CoDesign, 7*(3–4), 143–154. https://doi.org/10.1080/15710882.2011.630473

Mitra, S., Sætra, H. S., Raskin, J.-P., & Pansera, M. (2024). *Role of ICT innovation in perpetuating the myth of techno-solutionism.* https://doi.org/10.2139/ssrn.4907474

Monteiro, M. (2019). *Ruined by design: How designers destroyed the world, and what we can do to fix it.* Mule books.

Morozov, E. (2013). *To save everything, click here: The folly of technological solutionism.* PublicAffairs.

Nelson, H. G., & Stolterman, E. (2012). *The design way: Intentional change in an unpredictable world.* MIT Press. https://doi.org/10.7551/mitpress/9188.001.0001

Norman, D. (2024). *Design for a better world: Meaningful, sustainable, humanity centered.* MIT Press.

Olsson, T., Pyyhtinen, O., Rantasila, A., Laaksonen, S., Vigren, M., Ylipulli, J., & Sawhney, N. (2023). *A user-centered lens into digital excess: Exploring the superfluity and environmental burden of the digital world.* Computing within Limits. https://limits.pubpub.org/pub/krloqqv4/

Papanek, V. (1991). *Design for the real world: Human ecology and social change* (Revised 2nd ed.). Thames & Hudson.

Pierce, J. (2021). In tension with progression: Grasping the frictional tendencies of speculative, critical, and other alternative designs. In *Proceedings of CHI '21* (pp. 1–19). ACM. Article 617. https://doi.org/10.1145/3411764.3445406

Saleebey, D. (1994). Culture, theory, and narrative: The intersection of meanings in practice. *Social Work, 39*, 351–359.

Santarius, T., Bieser, J. C. T., Frick, V., Höjer, M., Gossen, M., Hilty, L. M., Kern, E., Pohl, J., Rohde, F., & Lange, S. (2023). Digital sufficiency: Conceptual considerations for ICTs on a finite planet. *Annals of Telecommunications, 78*, 277–295. https://doi.org/10.1007/s12243-022-00914-x

Sætra, H. S., & Selinger, E. (2023). *The siren song of technological remedies for social problems: Defining, demarcating, and evaluating techno-fixes and techno-solutionism.* https://doi.org/10.2139/ssrn.4576687

Schwarz, J. O., Wach, B., & Rohrbeck, R. (2023). How to anchor design thinking in the future: Empirical evidence on the usage of strategic foresight in design thinking projects. *Futures, 149*. https://doi.org/10.1016/j.futures.2023.103137

Selbst, A. D., Boyd, D., Friedler, S. A., Venkatasubramanian, S., & Vertesi, J. (2019). Fairness and abstraction in sociotechnical systems. In *Proceedings of the conference on fairness, accountability, and transparency.* ACM. https://doi.org/10.1145/3287560.3287598

Westlund, O. (2021). Advancing research into dark participation. *Media and Communication, 9*(1), 209–214. https://doi.org/10.17645/mac.v9i1.1770

Winner, L. (1980). Do artifacts have politics? *Dædalus, 109*(1), 121–136. http://www.jstor.org/stable/20024652

Winograd, T., & Flores, F. (1986). *Understanding computers and cognition: A new foundation for design.* Ablex.

Chapter 4

Calculating, Reducing, Compensating and Disconnecting: Summons of Digital Excess

Minna Vigren[a], *Tero Karppi*[b] *and Olli Pyyhtinen*[c]

[a]LUT University, Finland
[b]University of Toronto, Canada and LUT University, Finland
[c]Tampere University, Finland

Abstract

Digitalisation typically amounts to a modernist progress story of constant renewal, innovation and endless possibilities. The unprecedented scale of digital content production and storage has led to *digital excess*, with vast amounts of data, much of it redundant, accumulating without clear utility. Drawing from both Bataille's theory of excess and disconnection studies, the chapter conducts a situational analysis of how users are informed about the environmental consequences of digital excess and summoned to *calculate*, *reduce*, *compensate* and *disconnect* to mitigate their digital carbon footprint. While the digital carbon footprint is largely a structural problem, users are called on to engage in actions like reducing unnecessary data usage, adopting energy-efficient user practices, compensating emissions through carbon offset initiatives and detoxing from social media platforms. Overall, the authors argue that the concept of digital excess provides a useful tool for deconstructing different practices of dealing with the overflow of data.

Keywords: Digital excess; disconnection; digital carbon footprint; digital sustainability; environmental awareness; social media platforms

Introduction

'[W]hen it comes to data we have gone from talking about megabytes and gigabytes to talking about terabytes, petabytes, and exabytes. Our data use is exploding so much it feels like we keep building on the word byte', the host

Platforms and the Planet, 67–81

doi:10.1108/978-1-83662-172-020261005

Prerna Pant explains to the viewers in the documentary film *Greening the Net* (2023). Today, the unprecedented scale of the production and storage of digital content has led to *digital excess* (Olsson et al., 2023), with vast amounts of data, much of it redundant, accumulating without clear utility. This excessive and exponentially growing amount of data – estimated to reach 394 zettabytes by 2028 (Statista, 2024) – is polluting the planet in the form of energy consumption and greenhouse gas (GHG) emissions. In the documentary, Pant explores how we could possibly reduce our data usage, asking questions like whether it is better to use email instead of Facebook Messenger or SMS. The documentary ends in a lush Singaporean national park, where the greenery is vibrant and the ground's textures almost tangible. This ending nods to projects discussed in the film that offset the digital footprint of actions like ride-hailing or using search engines by planting trees. Pant is in the park, using a hoe to dig holes in the ground. After planting seven trees, she notes:

> I've had a productive morning. Usually I'd be taking pictures of this and sending it on WhatsApp, posting it on Facebook and Instagram, but today I am going to choose only one of those because from now on it's all about greening the net.

The choice to reduce and disconnect from social media platforms for the sake of the environment is underlined with a happy smile.

In our take on the overflow of data, we follow here Olsson et al. (2023), who consider the environmental burden of information and communication technology in terms of *digital excess*. Their understanding of digital excess draws from Georges Bataille (1988), a provocative French theorist who defines excess as the energy that cannot be used for growth but must be used, consumed and expended otherwise. As Bataille (1988, p. 21) insists, 'if the system can no longer grow, or if the excess cannot be completely absorbed in its growth, it must necessarily be lost without profit; it must be spent, willingly or not, gloriously or catastrophically'. Thus, the theory maintains that economic systems defined through use value find their limit in excess, when energy and production cannot be harnessed for utility (see also Groys, 2012). This excess manifests in things like luxury products and conspicuous consumption but also waste. For Bataille, the societal consequences of excess can even result in crises such as conflicts and wars (Groys, 2012, p. 119), but in the context of this chapter, we connect digital excess with the environmental crisis and climate change. Digital excess contributes to the digital carbon footprint and, by definition, amounts to something that is more than necessary, permitted or desirable. Olsson et al. adopt a 'user-centric' approach to digital excess, exploring it as part of 'subjectively perceived' aspects of digital life that do not 'carry meaningful value but, rather, appear as wasteful and superfluous, while also causing harm to individuals, society, or the planet' (Olsson et al., 2023, p. 2). The user-centric perspective is manifest also in *Greening the Net*. We see the documentary film not only as a fascinating example of the narratives surrounding the digital carbon footprint but also as a vignette of some of the key elements in the contemporary discussion of mitigating the digital carbon footprint and in the mediation of digital excess as an end-user problem. A self-described digital junkie, Pant is a mirror reflecting the viewer's own life and digital consumption habits: when looking at how she uses her digital devices to share photos, do video calls with family, work online and occasionally even take her phone with her to the washroom, the viewer cannot but feel a pang of guilt.

Contrary to the user-centric perspective, which underlines individual responsibility and the subjective perception of excess, our approach to digital excess could be called *user-decentring*. Instead of focusing primarily on the end user and their choices, our analysis highlights the ways in which end users are *summoned* – called on, invited, urged – by different stakeholders, such as climate protection organisations, governmental agencies, businesses and media, to take action and assume responsibility for mitigating the digital carbon footprint and its detrimental environmental impacts. We interpret these summons or calls as attempts to tackle users' alleged lack of awareness and willingness to act and reduce the carbon footprint associated with digital services like social media (Gnanasekaran et al., 2021). While corporations and their collect-it-all mentality have without doubt the largest share of responsibility for producing the excess of data (Crawford, 2021; see also Mahnke, this volume), in what follows, we provide an overview of how individuals are urged to assume responsibility for it in their everyday internet and social media use. With the notion of digital excess, we provide an analytical framework that can be used to deconstruct different practices of dealing with the overflow of data and subsequently to theorise what is at stake with these practices. Our intention is not to reproduce the neoliberal tendency to responsibilise individuals for solving societal problems (Miller & Rose, 2008; Rose, 1999) but rather to some extent also highlight the discrepancy between different scales in the solutions offered: while digital excess is first and foremost a structural problem, the solutions tend to emphasise the role of individual end users.

Methodology: Mapping Digital Excess

'[E]mpirical analysis of digital excess could help to recognize what in our digital lives might appear most irrelevant or unnecessary', Olsson et al. (2023, p. 2) maintain. They also observe the complexity of an individual's digital ecosystem fragmented into components like services, software, data, hardware and infrastructure, all of which enable the production and flow of information (Olsson et al., 2023). Mapping the irrelevant and unnecessary is not necessarily easy. Our empirical approach in this chapter follows loosely the method developed by Adele Clarke (2003) under the name of situational analysis. Clarke's lens is 'relational/ecological', drawing from thinkers like Deleuze and Guattari, Foucault and Strauss (Clarke, 2003; Fletcher & Clarke, 2018, p. 230). At the heart of this approach are the '*analytic exercises*' of mapping, which are intended to allow novel interpretations of the relations between actors in specific situations to emerge (Clarke, 2003, p. 560).

The mapping exercise allows us to establish connections between digital excess and the field of disconnection studies (see Bagger, 2024; Fast, 2021; Karppi, 2018; Light, 2014), which typically has focused on the reasons to quit and resist social and digital media. Moe and Madsen have identified a specific branch of disconnection studies with a focus on what is 'good for our climate and environment' (Moe & Madsen, 2021, p. 1595). While they are able to identify the framework, their examples are scarce. We contribute to this framework by mapping how digital excess, as a new digital episteme, turns into suggested concrete action plans to manage one's digital carbon footprint.

As a practice, mapping enables us to analyse different approaches to digital excess and their differential environmental relations. It allows us to see how different actors are situated within this landscape and how particular actors (including the platforms) are positioned in respect to one another and to digital excess in general. Moreover, it makes it possible for us to identify and reflect on emergent concepts and patterns in this specific discourse (cf. White & Marsh, 2006). The process of mapping helps us to understand how heterogeneous elements come together in the collected materials. This mapping identifies 'key elements of the situation, variation, and difference(s)', outlines the 'social worlds and arenas in mesolevel discursive negotiations' and establishes 'issues and discursive axes focused around difference(s) of positionality and relationality' (Clarke, 2003, pp. 558–559).

We are interested in what constitutes one particular situation: the moment where digital excess and awareness of the digital carbon footprint is increased and leveraged into summons to act. The mapping exercise was done by tracing online materials that discuss the mitigation of the individual digital carbon footprint. The data set includes over 30 websites and blog posts from climate protection organisations, governmental agencies and businesses, coupled with news pieces addressing the topic. The materials were almost entirely dated between 2021 and 2024, highlighting the emergent nature of raising consumer awareness on the topic. The online materials were qualitatively coded using NVivo, based on which of the four discursive tactics of *calculating, reducing, compensating* and *disconnecting* were identified, and illustrative examples for each were chosen.

Calculating

The climate impacts of digital media use have entered public discussion through a range of alarming figures. Exact quantification of the environmental impact of our digital lives has, however, proven challenging (e.g., Laaksonen & Premsankar, this volume; Pasek et al., 2023). Yet, numbers and comparisons play a key role in making the digital carbon footprint knowable and manageable for individual internet users. Metrics are believed to be endowed with power to increase individual awareness and offer a starting point to reflect and change one's own behaviour. Jorge et al. (2022) have shown that in response to the techlash, big tech companies introduced digital tools to measure and display users' individual screen time. Big tech's proposition was that these tools would create more self-awareness and allow users to better optimise their social media engagements.

A similar trope is visible in the digital carbon footprint discourse as well. One digital carbon footprint calculator is offered by Australian comparison service provider Compare the Market (n.d.). The calculator is based on the environmental footprint metrics for social media applications by Greenspector, a French company specialising in 'eco-responsible digital transformation' (Greenspector, n.d.). Compare the Market introduces The Social Carbon Footprint Calculator by inviting individuals to reflect on the importance of social media sites in connecting us to our friends and family and keeping us up to date with what is going on in the world. The company reminds about the environmental impact of social media sites (Greenspector, n.d.):

> Just like all the other technology that we use, they have an environmental footprint too. And while a quick scroll through your feed or the odd YouTube video isn't going to use up too much energy, over the course of a year things can start to add up, especially when you consider just how many people use these social media sites across the world.

The calculator functions so that, by inserting an estimation of how many minutes one spends on 10 of the most popular social media sites per day, the user is delivered CO_2 calculations for each platform per day, per year and as a total for all platforms. Whereas Compare the Market refrains from comparing the numbers to the carbon footprint of other aspects of life, others have sought to make them more relatable for individual end users. For example, Carbon Literacy Project (Walkley, 2023) and World Clean Up Day (Komazova, 2024) refer to the Greenspector's calculations comparing the CO_2 emissions of social media platforms to those of driving a car, a frequently cited culprit of climate change. Furthermore, Compare the Market's calculator is not accompanied with advice for the individual on how to reduce one's digital carbon footprint, thus leaving the user without actionable guidance.

Media scholars Marks and Przedpełski (2022, p. 213) have examined different efforts to calculate the digital carbon footprint of streaming videos. The authors note that there is an enormous disparity between figures in the literature and 'a surprising degree of cherry-picking when it came to identifying data, modelling electricity consumption, and prediction'. In alignment with their observation, we argue that the power of calculations to determine the digital carbon footprint seems to lie less in their truth value – that is, in how the number corresponds to the actual emissions – and more in how people are accustomed to associating numbers with objectivity, truth and accountability: you first need a number to make something credible and then act upon it. 'The credibility of numbers, or indeed of knowledge in any form, is a social and moral problem', historian of science Porter (1995, p. 11) suggests. He implies that our 'faith in objectivity' is exploited when bureaucrats and decision-makers appeal to numbers, for a 'decision made by the numbers (or by explicit rules of some other sort) has at least the appearance of being fair and impersonal' (Porter, 1995, p. 8). This observation is corroborated by Marks and Przedpełski (2022, pp. 213–214), who note that the mismatches of numbers seem to reflect not only the regular turnover of scientific findings but also ideological agendas, which highlights the need for a critical interrogation of digital carbon footprint calculations, not as neutral metrics of environmental impact but as socially constructed tools that can obscure power dynamics and serve ideological purposes under the guise of objectivity.

Reducing

'Here's another reason to get off your device! Every click, tap, and download contributes to your digital carbon footprint', instructs a text on the website of the David Suzuki Foundation (n.d.-b), a Canadian non-profit organisation. Since the year 1990, it has set out to 'protect nature's diversity and the well-being of all life, now and for the future', and spreads a vision 'that we all act every day on the understanding that we are one with nature' (David Suzuki Foundation n.d.-a). According to its own statement, the foundation is Canada's 'most credible and reliable source of evidence-based environmental information' (David Suzuki Foundation n.d.-a).

Over the past few years, a range of actors – including the European Union, UNICEF, Ofcom, various climate action initiatives and eco-companies – have addressed the individual to become more mindful about the environmental costs of their daily digital practices. The David Suzuki Foundation provides one of the most comprehensive lists on how to reduce the digital carbon footprint (see also e.g., Netzero, 2024; Ofcom, 2023; UNICEF, 2023). One can begin with small adjustments, such as using the dark mode and power-saving settings, turning off devices and using WLAN rather than mobile networks. Further action is encouraged to change mundane habits by avoiding sending unnecessary emails and attachments, opting for lower resolution settings when streaming, downloading content instead of streaming, using less artificial intelligence and closing tabs in one's browser (David Suzuki Foundation, n.d.-b).

The excess of digital daily life is most directly addressed in tips for decluttering (David Suzuki Foundation, n.d.-b). These include deleting unused files and apps, being mindful of downloading new apps, clearing web browsers' cache and cookies, as well as deleting duplicates and unnecessary digital bookmarks from the web browser. The Foundation emphasises the energy implication of digital decluttering, pointing to the responsibility of the individual:

> Every file, app or program stored on your device requires energy to create, maintain and access. The more cluttered your device is, the more energy it consumes during use. And if you use a cloud server, your data is syncing regularly to data centers and increasing their energy demand. (David Suzuki Foundation, n.d.-b)

The practices of reduction are strategies of mitigation that do not require total abstention. Reduction can be about a choice between which platform one uses for communication (instead of all of them). It can also be about actively choosing low-resolution content rather than always aiming for high-fidelity. Even though the reduction narrative is appealing in its simplicity, its effectiveness has been challenged. For example, Marks and Przedpełski (2022, p. 213) have examined different attempts to work with the digital carbon footprint in relation to streaming videos both in theory and practice. Marks' and her colleagues' Small File Media Festival (2024) is an example of the latter; the films screened at the festival must adhere to strict low-resolution limits, streaming at only 1.44 megabytes per minute, thus significantly reducing electricity usage compared to the high-definition standards of contemporary streaming media. While the festival is organised in terms of data reduction, Marks and Przedpełski (2022, p. 214) also point towards a specific problem in the approach: 'networks and data centers are running 24/7, regardless of data use' (while using less energy, though, when idle). Individual attempts to reduce data use once again face the larger structural problem – even the data centres are created for future potential, expansion and excess, rather than actual data use. For Marks and Przedpeiski, it is not the individual use of devices that creates the problem but the expected expansion of this use in the future. 'The more we use them, the more the infrastructure will expand. *That* is why streaming is responsible for an increase in ICT's carbon footprint', they write and continue, 'Our goal can only be the equivalent of keeping more planes out of the sky: reducing the *expansion* of ICT' (Marks & Przedpełski, 2022, p. 215).

Compensating

Compensation is yet another solution offered for mitigating the harms of platform capitalism. Of the services available, the Danish start-up company EcoTree provides an illuminating example. EcoTree promotes 'environmental awareness and engaging individuals and businesses in eco-friendly behaviour' (The Hub, n.d.). On their website, the company provides calculators to estimate carbon emissions from flights, cars, trains, homes and internet use, revealing the infrastructures that move goods, people and data. In terms of digital platforms, the metrics related to the CO_2 emissions of internet use are of our interest here. The metrics are followed by tips for reducing the internet-related CO_2 emissions, in line with the suggested actions of the David Suzuki Foundation and others. Moreover, EcoTree declares: 'We don't have a magic wand, but we do have the next best thing!' (EcoTree, n.d.-a). By this they mean web shop for mitigating residual emissions by buying trees and carbon credits. Individuals are urged to 'earn a clean conscience' (EcoTree, n.d.-a) by buying trees to support 'the creation of carbon sinks through the maintenance and renewal of forests and their ecosystems' (EcoTree, n.d.-b). The customer is advised to browse the trees and 'choose the one that feels right', because 'after all, it's going to be part of your life for a long time' (EcoTree, n.d.-c). The trees can be bought for oneself or as a gift, and there are multiple options available, including, for example, a young Red Cedar in Cléden-Poher 2 Forest (France) for 18 € or a little French forest of 30 trees available in different regions for 552 €. EcoTree gives the tree buyer a chance to 'participate in financing sustainable forest management handled by professionals and in contributing to the development of a forest, while getting a possible return on their investment when the purchased trees reach their harvesting date' (EcoTree, 2021).

For those who wish to 'contribute more' (EcoTree, n.d.-c), there is a possibility to purchase carbon credits, one carbon credit being equivalent to capturing one ton of CO_2. For example, buying one carbon credit in Nouïc Forest in Nouvelle-Aquitaine (France) costs 72€. The locations in Europe are chosen in forestry projects 'that meet the criteria of uniqueness, additionality, permanence and measurability', to enable CO_2 emissions capturing 'as locally as possible' in Europe (EcoTree, n.d.-d).

Carbon offsetting is about trading 'rights' or certificates to compensate the buyer's CO_2 emissions with an equal amount of CO_2 reductions somewhere else. The benefits of offsetting have faced widespread criticism due to evidence of inaccurate and misleading false crediting (Greenfied, 2023), reports of human rights violations against Indigenous communities (Human Rights Watch, 2024) and concerns that the system enables companies and countries to evade direct reductions in their CO_2 emissions (Cullenward et al., 2023). In its carbon manifesto, EcoTree acknowledges these problems, emphasises the necessity for global-scale solutions and rejects the notion of achieving carbon neutrality solely at the individual level (EcoTree, n.d.-b). Hence, carbon footprint calculators and a carbon credit webshop for individuals appear contradictory. To address this potential critique, EcoTree (n.d.-b) stresses that offsetting is secondary to their main principle, avoiding and reducing. Yet they advocate carbon crediting as well because 'efforts to reduce emissions alone will not be enough to reverse the trend towards climate disruption. That's why it's right to take all these actions at the same time' (EcoTree, n.d.-d).

To bring this back to Bataille's (1988, p. 29) theory of excess, our own planet, the biosphere, 'which corresponds to the space available to life, is the only real limit'. In the compensation narrative, the excessive expansion of platform capitalism, quite physically in the form of data centres, takes space from the living (humans, nature, animals). Planting trees becomes a way of expanding or renegotiating these limits. Trees, as we know, absorb carbon dioxide from the atmosphere and are vital for life. Yet, the forest behind the trees is neoliberal capitalism, which prioritises individual action over systemic change. Planting trees or purchasing carbon credits underscores the tension between the appeal of accessible, market-based solutions and the urgent need for collective, structural transformations to address the root causes of environmental crises.

Disconnecting

One solution for reducing one's carbon footprint by the David Suzuki Foundation is simple: 'Take a break from your devices! Not only is this good for your mental health, it also reduces your digital carbon footprint' (David Suzuki Foundation, n.d.-b). The Foundation deconstructs the idea of digital detox into smaller action items, familiar from research and self-help literature on disconnection (e.g., Goodin, 2017; Syvertsen & Enli, 2020; Zahariades, 2018). The list on the website (David Suzuki Foundation, n.d.-b) includes:

> Regularly schedule digital downtime in your day. Try not using devices during meals, before bedtime or during family and friend time. Set screen time limits on streaming and social media apps. Designate certain areas in your home, such as the bedroom or dining room, as tech-free zones. Limit unnecessary notifications to minimise distractions and temptations to constantly check your phone or computer. Disconnect from your device and reconnect with nature. When you have the impulse to scroll, go outdoors and stroll.

Another example of a similar summons to abstention is the Green Sabbath Project (n.d.), which states on its website: 'Is there *nothing* you can do about the environment? *Nothing* may be one of the best things you can do. One day every week. Do *nothing.*' When all else fails, not doing or doing nothing is perhaps the last solution with which an individual user is left.

The practices of withdrawing and doing nothing, of course, have a history that goes well beyond the age of social and digital media. The founder of the Green Sabbath Project, Jonathan Schorsch (2022), Professor of Jewish Religious and Intellectual History, has suggested the practice and ritual of the sabbath as the way to mitigate the environmental crisis caused by hypercapitalism. Like the David Suzuki Foundation, The Green Sabbath Project tries to raise awareness and educate people, but the project's inspiration comes from the traditional principles and practices of Jewish and Christian sabbath that are centred on '(a) rest, withdrawal, doing "nothing"; and (b) regeneration, renewal and repair' (Schorsch, 2022, p. 237).

The ritual of sabbath proposed by the project can be either secular or spiritual. One of the examples on the project website is 'The Offline Club', which regularly meets in a café established in an old church in the city of Amsterdam in the Netherlands (Karakus, 2024). The club is inspired by the idea of digital detox. As Özlem Karakus (2024) describes it, 'the club encourages participants to lock away their smartphones and immerse themselves in a series of offline activities ranging from board games to deep conversations, all set in an environment designed to foster relaxation and creativity.'

The Offline Club, but also rest, withdrawal and doing nothing more generally, can in this context be described as 'disconnective practices' (Beattie & Cassidy, 2021; Light, 2014; Mannell, 2017). Ben Light (2014, p. 150) notes that this focus is 'not about just one practice to fit all scenarios, but is a suite of activity involving a variety of people and things'. To be clear, disconnective practices should not be seen merely as practical choices. As Cascone and Bonini (2024, p. 4) write, disconnection is 'a critical, political and conscious response to the excesses of the culture of connectivity'.

But there is no rest for the wicked – the user, the influencer, the small business owner, whose capital, social and economic alike, is tied to digital platforms. For them, disconnection is an extreme choice. Internet scholar Alice Marwick (2023, pp. 71, 86) calls it the 'nuclear option'. For Schorsch (2022, p. 241) of Green Sabbath, it is abstention in a 'maximal' form. Disconnection is extreme because in a 'platform society', connectivity is intimately tied with modern life (van Dijck et al., 2018). Therefore, abstaining from one's online actions means cutting off and detaching oneself from functions that are integral to contemporary society. In the context of digital carbon footprint, the connection between digital media and the forms and powers of contemporary society is rendered visible. Some have defined disconnection as a form of 'digital suicide' (Karppi, 2018), but we could also think of it in terms of *sacrifice,* connecting it with how Bataille describes dealing with excess. Bataille's (1988, pp. 67–68) example is 'potlatch', an agonistic Indigenous gift-giving ceremony. Potlatch offers an example of gift-giving in the form of sacrifice – the giver voluntarily lets go of their possessions to deal with the detrimental impacts of excess. Interestingly, this logic also seems to apply to the practice of disconnection. The scale and extent of this sacrifice range from the fear of missing out to not being able to take part in the services of modern society and its daily functions. An important aspect for Bataille (1988, p. 67) is that the sacrifice is not an individual but a social act. Thus, in sacrifice, one does not only lose but also gains. When you publicly de-centre yourself from your previous position of wealth in the eyes of others, you structure your position in the system in a new way and, indeed, maybe become part of its change. When disconnection is understood as a sacrifice rather than a solution or cure, the moral and ethical boundaries of connectivity are thereby rendered more visible.

Conclusion: A User-Decentring Lens on Digital Excess

In this chapter, we examined the ways in which users are summoned to consider and manage digital excess and the related environmental harms from using the internet and platforms in online materials. We introduced four key practices – *calculating, reducing, compensating* and *disconnecting* – that are presented as means not only to make digital excess more knowable and manageable but also to reduce one's digital footprint. All of them frame the end user as the principal actor whose responsibility it ultimately is to deal with the resulting digital excess. The actions of managing and reducing one's digital excess can be interpreted, following social media scholar Niall Docherty (2021, p. 3829), as a 'responsibilized construct of neoliberal well-being', configuring individuals as agents of self-discipline and habit control with the obligation to mitigate their digital carbon footprint.

In contrast to the user-centred lens of Olsson et al. (2023) addressing how users' actions produce digital excess, the *user-decentring* approach that we suggested in this chapter contextualises the user-centred technological and moral fixes to digital excess, which in our view only touch on the symptoms without getting at the root of the problem. Digital excess is first and foremost a structural problem, but placing the responsibility on the individuals tends to hide this. It is companies that are largely responsible for producing digital excess, and digital excess is also an integral part of the operations and chosen order of platform companies, for example (see also Laaksonen & Premsankar, this volume). Social media users are invited to produce digital excess, as data in general – rather than in particular – are deemed profitable by the companies that own and govern the platforms.

This is where the stakes of our approach lie: troubling the user-centred solutions and contextualising them in wider systems. Yet, there is a potential for change in the summons to mitigate digital excess themselves. To summon is to call upon the individual user to be present and to manage their own digital excess – a call that can readily be interpreted through the framework of self-responsibilisation, wherein individuals internalise their position as particular neoliberal user-subjects. Yet, in his study on 'user-de-centeredness design', HCI scholar Yutaka Yamauchi (2016, p. 591) suggests that, rather than being assigned a ready-made role, users can also be 'challenged and drawn into a struggle, and as a result are encouraged to treat their own subjectivity as problematic and subsequently to make efforts to transform it'. In other words, to summon is also to evoke or call forth powers that do more than assign an individual to an established social role, for power is relational, and can not only constitute but also undo normative forms of user subjectivity. The summons to mitigate digital excess may therefore – at least potentially – be also 'transgressive' practices, which for Bataille (1986) are important acts of resistance for they allow opening the structures that restrict our being, existence and freedom. Our analysis highlighted the processes through which users are called on and drawn into reflecting and problematising their actions. In those processes, users' positions, responses and responsibilities are continuously negotiated and never emanate solely from the individual subject but from within the system that poses its limits for us and is challenged. This is where the change can begin: calculating, reducing, compensating and disconnecting have the potential for breaking the established norms and boundaries of digital connectivity – boundaries that are defined by the platform companies and can be redefined by us, if we realise that they are artificial constructs and not set in stone.

Acknowledgements

The work of the first author was funded by the *Imagining Sustainable Digital Futures* project, Research Council of Finland [grant number: 347950]. The work of the third author was financially supported by the European Union [ERC, WasteMatters, grant number: 101043572] and the Research Council of Finland [grant number: 350191]. Views and opinions expressed are those of the author only and do not necessarily reflect those of the European Union or the European Research Council Executive Agency or any other funder named. Neither the European Union nor the granting authorities can be held responsible for them.

References

Bagger, C. (2024). A decade of digital disconnection research in review: Where, what, how, and who? In K. Albris, K. Fast, F. Karlsen, A. Kaun, S. Lomborg, & T. Syvertsen (Eds.), *The digital backlash and the paradoxes of disconnection* (pp. 109–128). Nordicom, University of Gothenburg. https://doi.org/10.48335/9789188855961

Bataille, G. (1986). *Erotism: Death & sensuality* (1st City Lights ed.). City Lights Books.

Bataille, G. (1988). *The accursed share: An essay on general economy. Volume I Consumption*. Zone Books.

Beattie, A., & Cassidy, E. (2021). Locative disconnection: The use of location-based technologies to make disconnection easier, enforceable, and exclusive. *Convergence, 27*(2), 395–413. https://doi.org/10.1177/1354856520956854

Cascone, M., & Bonini, T. (2024). "Disconnecting from my smartphone is a privilege I do not have": Mobile connection and disconnection practices among migrants and asylum seekers in three migrant reception centres of Sicily. *New Media & Society*, *27*(9), 5201–5219. https://doi.org/10.1177/14614448241249371

Clarke, A. (2003). Situational analyses: Grounded theory mapping after the postmodern turn. *Symbolic Interaction*, *26*(4), 553–576.

Compare the Market. (n.d.). *Social carbon footprint calculator*. https://www.comparethemarket.com.au/energy/features/social-carbon-footprint-calculator/

Crawford, K. (2021). *Atlas of AI: Power, politics, and the planetary costs of artificial intelligence*. Yale University Press.

Cullenward, D., Badgley, G., & Chay, F. (2023). Carbon offsets are incompatible with the Paris Agreement. *One Earth*, *6*(9), 1085–1088.

David Suzuki Foundation. (n.d.-a). *About the David Suzuki Foundation*. https://davidsuzuki.org/about/. Accessed on November 13, 2024.

David Suzuki Foundation. (n.d.-b). *How to reduce your digital carbon footprint*. https://davidsuzuki.org/living-green/how-to-reduce-your-digital-carbon-footprint/. Accessed on November 13, 2024.

Docherty, N. (2021). Digital self-control and the neoliberalization of social media well-being. *International Journal of Communication*, *15*, 3827–3846. https://ijoc.org/index.php/ijoc/article/view/15272

EcoTree. (2021). *EcoTree*. https://cdn.ecotree.green/documents/EcoTree_Notice%20of%20Information_Sep21_EN.pdf. Accessed on December 11, 2024.

EcoTree. (n.d.-a). *Calculate and reduce your internet's CO_2 emissions*. https://ecotree.green/en/calculate-digital-co2. Accessed on August 20, 2024.

EcoTree. (n.d.-b). *Our carbon manifesto*. https://ecotree.green/en/carbon-offset-manifesto. Accessed on November 13, 2024.

EcoTree. (n.d.-c). *Welcome to the TreeShop*. https://ecotree.green/en/treeshop. Accessed on August 20, 2024.

EcoTree. (n.d.-d). *Contribute to the fight against climate change: Buy high-quality carbon credits*. https://ecotree.green/en/carbon-credit-shop. Accessed on August 20, 2024.

Fast, K. (2021). The disconnection turn: Three facets of disconnective work in post-digital capitalism. *Convergence*, *27*(6), 1615–1630. https://doi.org/10.1177/13548565211033382

Fletcher, I., & Clarke, A. E. (2018). Imagining alternative and better worlds: Isabel Fletcher talks with Adele E. Clarke. *Engaging Science, Technology, and Society*, *4*, 222–245. https://doi.org/10.17351/ests2018.216

Gnanasekaran, V., Fridtun, H. T., Hatlen, H., Langøy, M. M., Syrstad, A., Subramanian, S., & De Moor, K. (2021). Digital carbon footprint awareness among digital natives: An exploratory study. *NIKT*, *1*, 99–112. https://www.ntnu.no/ojs/index.php/nikt/article/view/5226/4702

Goodin, T. (2017). *Off: Your digital detox for a better life*. Abrams.

Green Sabbath Project. (n.d.). *Green Sabbath Project*. https://www.greensabbathproject.net. Accessed on November 13, 2024.

Greenfield, P. (2023, January 18). Revealed: More than 90% of rainforest carbon offsets by biggest certifier are worthless, analysis shows. *The Guardian*. https://www.theguardian.com/environment/2023/jan/18/revealed-forest-carbon-offsets-biggest-provider-worthless-verra-aoe

Greening the Net. (2023). *A documentary film produced by CNA Insider*. https://www.youtube.com/watch?v=eOPetDY-nss

Greenspector. (n.d.). *About Greenspector*. https://greenspector.com/en/enterprise/. Accessed on August 19, 2024.

Groys, B. (2012). *Under suspicion: A phenomenology of media*. Columbia University Press.

Human Rights Watch. (2024). *Carbon offsetting's casualties: Violations of Chong indigenous people's rights in Cambodia's Southern Cardamom REDD+ project*. https://www.hrw.org/sites/default/files/media_2024/02/cambodia0224web_1.pdf

Jorge, A., Amaral, I., & Alves, A. (2022). "Time well spent": The ideology of temporal disconnection as a means for digital well-being. *International Journal of Communication*, *16*, 1551–1572. https://ijoc.org/index.php/ijoc/article/view/18148/3717

Karakus, Ö. (2024, July 4). *The Offline Club: Reconnecting in a disconnected world*. https://www.greensabbathproject.net/all-posts/the-offline-club-reconnecting-in-a-disconnected-world

Karppi, T. (2018). *Disconnect: Facebook's affective bonds*. University of Minnesota Press.

Komazova, I. (2024, January 15). *Digital decluttering: A sustainable approach to social media use*. https://www.worldcleanupday.org/post/digital-decluttering-a-sustainable-approach-to-social-media-use

Laaksonen, S.-M., & Premsankar, G. (this volume). The challenging task of estimating the carbon footprint of digital services.

Light, B. (2014). *Disconnecting with social networking sites*. Palgrave Macmillan.

Mannell, K. (2017). Technology resistance and de Certeau: Deceptive texting as a tactic of everyday life. *Platform: Journal of Media & Communication*, *8*(1), 40–55.

Marks, L. U., & Przedpełski, R. (2022). The carbon footprint of streaming media: Problems, calculations, solutions. In P. Kääpä & H. Vaughan (Eds.), *Film and television production in the age of climate crisis. Palgrave Studies in Media and Environmental Communication* (pp. 173–188). Palgrave Macmillan. https://doi.org/10.1007/978-3-030-98120-4_10

Marwick, A. E. (2023). *The private is political: Networked privacy and social media.* Yale University Press. https://doi.org/10.12987/9780300271652

Miller, P., & Rose, N. (2008). *Governing the present: Administering economic, social and personal life.* Polity Press.

Moe, H., & Madsen, O. J. (2021). Understanding digital disconnection beyond media studies. *Convergence, 27*(6), 1584–1598. https://doi.org/10.1177/13548565211048969

Netzero. (2024, April 9). *12 ways to reduce your digital carbon footprint.* https://www.netzero.im/latest/12-ways-to-reduce-your-digital-carbon-footprint/

Ofcom. (2023, July 5). *Five ways to help reduce your digital carbon footprint.* https://www.ofcom.org.uk/internet-based-services/technology/reduce-your-digital-carbon-footprint/

Olsson, T., Pyyhtinen, O., Laaksonen, S.-M., Vigren, M., Ylipulli, J., Rantasila, A., & Sawhney, N. (2023). A user-centered lens into digital excess: Exploring the superfluity and environmental burden of the digital world. In *Proceedings of the LIMIT ninth computing within limits 2023.* https://doi.org/10.21428/bf6fb269.a5916a92

Pasek, A., Vaughan, H., & Starosielski, N. (2023). The world wide web of carbon: Toward a relational footprinting of information and communications technology's climate impacts. *Big Data & Society, 10*(1), 1–14. https://doi.org/10.1177/20539517231158994

Porter, T. M. (1995). *Trust in numbers: The pursuit of objectivity in science and public life.* Princeton University Press.

Rose, N. (1999). *Powers of freedom: Reframing political thought.* Cambridge University Press.

Scorsch, J. (2022). Green Sabbaths: Putting the Anthropocene era to rest (once a week). In R. Borde, A. A. Ormsby, S. M. Awoyemi, & A. Gosler (Eds.), *Religion and nature conservation: Global case studies* (pp. 195–212). Routledge. https://doi.org/10.4324/9781003246190

Small File Media Festival. (2024, October 21–27). https://smallfile.ca/about/

Statista. (2024). *Volume of data/information created, captured, copied, and consumed worldwide from 2010 to 2023, with forecasts from 2024 to 2028.* https://www.statista.com/statistics/871513/worldwide-data-created/

Syvertsen, T., & Enli, G. (2020). Digital detox: Media resistance and the promise of authenticity. *Convergence, 26*(5–6), 1269–1283. https://doi.org/10.1177/1354856519847325

The Hub. (n.d.). *EcoTree.* https://thehub.io/startups/ecotree. Accessed on December 11, 2024.

UNICEF. (2023, April 17). *Five tips for reducing your digital footprint.* https://www.unicef.org/armenia/en/stories/five-tips-reducing-your-digital-footprint

van Dijck, J., Poell, T., & de Waal, M. (2018). *The platform society: Public values in a connective world.* Oxford University Press.

Walkley, S. (2023, March). *The carbon cost of social media.* https://carbonliteracy.com/the-carbon-cost-of-social-media/

White, M. D., & Marsh, E. E. (2006). Content analysis: A flexible methodology. *Library Trends*, *55*(1), 22–45. https://doi.org/10.1353/lib.2006.0053

Yamauchi, Y. (2016). User-de-centeredness in service design. In T. Maeno, Y. Sawatani, & T. Hara (Eds.), *Serviceology for designing the future (ICServ 2014). Selected and edited papers of the 2nd international conference on serviceology*. Springer. https://doi.org/10.1007/978-4-431-55861-3_40

Zahariades, D. (2018). *Digital detox: The ultimate guide to beating technology addiction, cultivating mindfulness, and enjoying more creativity, inspiration, and balance in your life*. An ArtOfProductivity.com Action Guide.

Section 2

Infrastructures and Materiality

Chapter 5

The Archival Paradox: DNA-Based Data Storage and the Data Hoarding Crisis

Mél Hogan[a], *Deb Verhoeven*[b] *and Tessa J. Brown*[a]

[a]Queen's University, Canada
[b]University of Alberta, Canada

Abstract

Awareness of the environmental impact of big tech, and in particular of massive, energy-intensive data centres worldwide, is growing rapidly. Given the reticence of Big Tech firms to accurately report their carbon emissions, it is difficult to provide an exact figure, but we know they are more than 7 times greater than these firms claim, and likely growing further when we include activities like bitcoin mining and the rapid rise of AI. Tech companies and start-ups are seeking less environmentally damaging data-storage solutions. Among the most promising are deoxyribonucleic acid (DNA)-based data storage and computing – those developing this technology suggest DNA has the potential to replace our current, unsustainable data storage systems. In this piece, the authors consider DNA-based data storage from three perspectives: first, the authors explore the technologies being developed, accounting for the environmental toll of current storage modalities. What kinds of materials and processes are involved in the production of synthetic DNA, and in using it to store and extract data? Second, the authors explore discourses around this developing technology, looking at how the companies behind this tech talk about and conceptualise the work they are engaged in, and how it is framed in the larger cultural context. Finally, the authors approach the subject through a more critical lens, considering whether transferring the vast stores of data currently held by tech giants like Amazon, Google and Meta to more 'sustainable' systems is likely to solve the problem of the ever-expanding demand for data storage, or whether a deeper shift in our attitudes towards data and technology is needed to safeguard our future and the future of our planet.

Platforms and the Planet, 85–97

doi:10.1108/978-1-83662-172-020261006

Keywords: DNA data storage; data centres; archival paradox; data hoarding; cloud technologies; big tech and the environment

> Something has changed inside us, as a house changes when a guest comes into it. We cannot say who has entered, we may never know, but there are many indications that the future enters us in just this way, to transform itself within us long before it happens. - Rainer Maria Rilke, 1902.

Introduction: The Data Hoarding Crisis and the Archival Paradox

Data and the appetite to stow it are ramifying. Once the specialist concern of national archives and institutional repositories, data saving is now ritually practiced at the most personal level. Fuelled by the shameful prospect of a spinning wheel of death, we intone the mantra of the backup. Backup again. Find the redundancy. Aim to be flawless. Focus on life expectancy. Be always plugged in. As our personal data hoarding reaches fever pitch, we unwittingly supply the colonising avarice of tech giants greedy for ever more digital real estate. Big tech is reaping the profits that come from our unclenching grip on the data under our fingertips. Add to this the ravenous data storage demands of artificial intelligence (AI). It all makes for a data centre earnings bonanza and an environmental disaster.

Data hoarding is a global problem, with 50–90% of the data most companies store classified as 'dark data' (Hawker, 2023). Dark data refer to the vast amounts of information collected and stored by organisations that are rarely, if ever, used or analysed. These repositories of accumulated but now forgotten information are also the fodder that feeds AI. Moreover, with a little over 57% of all new content on the internet now AI-generated (Harrison Dupré, 2024), AI is further transforming dark data by generating massive volumes of new, often purposeless stored information. This threatens to become the ultimate form of masticatory obsolescence – troves of archived information created without inherent value, constantly chewed and eschewed and all the while consuming digital and physical resources.

All these data are no longer kept 'in-house' but have headed to 'the cloud', a physical location that stores multitudes of hard drives and related technical infrastructure. Cloud computing, a concept which emerged between the mid-1990s and the mid-2000s (Regalado, 2011), has become ubiquitous over the past 20 years. Louise Amoore (2018) suggests that the metaphor of the cloud describes 'the advent of processes at scales that appear to transcend the observational paradigm and exceed our capacities to see and to understand' (p. 5). Conveniently out of sight, the sky-high terminology of the cloud serves to obscure the direct connection between personal or even organisational information hoarding from the data infrastructures that gird them (Mosco, 2015; see also Bartlett & Velkova, 2026, this volume).

This decoupling of everyday information practices from information infrastructure has shaped understanding and strategies for redressing the materially damaging impact of the cloud, specifically in the form of data centres. In storing our present data for posterity, data centres are at one and the same time contributing directly to the demise of the physical environment that sustains them. This, we contend, is the archival paradox. Specifically, the archival paradox refers to the contradictory logics underpinning contemporary data preservation: in attempting to secure our digital present for the future, we accelerate the destruction of the material conditions that make that future possible. Data centres, storage media and computational infrastructures consume vast ecological resources, emitting carbon and generating toxic waste, even as they promise permanence and security. At the same time, this act of preservation is never neutral. What gets archived – and what does not – is shaped by systems of power, exclusion and control, reinforcing dominant narratives while erasing others. The archival paradox is thus twofold: materially unsustainable and symbolically violent, it enshrines selected fragments of the present at the cost of both environmental survival and epistemic justice.

To date, the response to this existential archival conundrum has been limited to either: disavowal and disinformation (what problem? the numbers aren't as bad as you think); the use of lower-carbon energy alternatives and carbon offsets (fire up the nuclear reactor!); national regulatory interventions often without accountability mechanisms; tech-solutionism (AI will only get smarter, storage will only become more efficient). Commentators have been quick to point out the limitations of these various approaches. Luccioni et al. (2025), following Jevons but specifically writing about the environmental impact of AI, note that tech efficiency gains typically result in increased consumption (see also Mahnke, 2026, this volume). We would add that socially shared beliefs, expectations and practices around data storage are rarely examined as part of the problem – or solution – to the environmental impact of data storage.

If data centres are fashioned as a refurbished, digital version of the traditional archive, they are not just aiming to serve our present data needs but also 'preserve' them. Figured as information bunkers, data centres envisage a way for 'humanity' (if not humans) to survive an apocalypse, proffering immortality in the form of a hyperscale time capsule (Mendell et al., 2022). In this framework, we are not so much directing and shaping the future through the overabundant collection of digital keepsakes, but rather, as Rilke precipitously noted in 1902, the future (in the sense that we may not have one) has already transformed us and our storage practices (Rilke, 2012).

In this chapter, we use the advent of a new data storage technology based on synthetic deoxyribonucleic acid (DNA) to explore the archival paradox. Our focus is directed to studying the emerging professional settings that have developed to enable and mediate this technology. We adopt a reflection-in-action framework to identify how complexity, instability and uncertainty are intrinsic to contemporary professional knowledge, especially those defined by multiple and sometimes competing or contradictory practices (Schön, 1983, pp. 17–19). In keeping with this approach to DNA-based data storage, we draw on various sites of research including dispersed information resources (company websites, scholarly articles, news reports and promotional materials) as well as observation and site visits to Catalog DNA in the Boston area and the University of Washington, which, in partnership with Microsoft, houses the first fully automated DNA-based data storage device. We also interviewed, via Zoom, experts from multiple fields and institutional settings to expand our understanding of this emerging field beyond the hype generated from within it. Interviewees were selected to represent a wide range of geographic locations and perspectives on DNA-based storage and included scientists, artists, archivists as well as the public relations directors at key commercial enterprises, including Twist Bioscience, Digital Bedrock and Biomemory. Together, these various consultational approaches yielded the history, analysis and interpretation of the emergence of DNA storage technologies we present in this article.

Data Centres and the Environment

Many commentators have noted how the language of the cloud has served to conceal its environmental impact by evoking the ephemeral, the nebulous (or perhaps cirrus or cumulonimbus), rather than a concrete, physical system (Burrell, 2020; Carruth, 2014; Cubitt et al., 2011; Hogan, 2016; Hu, 2015; Mosco, 2015). However, more recently, awareness of the massive data centres and their steadily increasing carbon footprint has begun to spread beyond a small number of climate-conscious researchers.

A single hyperscale data centre can use as much power as 80,000 (average US) households. McKinsey has estimated that in the US market alone (40% of the global market), power consumption is expected to reach 35 gigawatts (GW) by 2030, up from 17 GW in 2022 (McKinsey Consulting, 2023). Recent analysis found that even this alarming prediction is likely a gross underestimate of actual data centre power consumption. Energy usage by big tech firms (Google, Microsoft, Meta and Apple) is calculated to be more than 650% higher than officially reported (O'Brien, 2024, see also Laaksonen & Premsankar, 2026, this volume).

In accordance with increasing day-to-day reliance on retrievable digital data, exacerbated by the rapid commercial uptake of generative AI, there is a corresponding increase in sustainability pressure on data centres. Regulators and governments such as Singapore and the Netherlands have introduced sustainability standards for newly built data centres, which promise improvements in power usage effectiveness (PUE), the amount of power a data centre uses relative to its total energy consumption. Still, after showing some initial progress, PUE has flattened over the past decade and has not offset the rising environmental impact of data centres (Uptime Institute, 2024).

Around 40% of a data centre's energy is devoted to cooling (McKinsey Consulting, 2023). Under commercial and political pressure to meet sustainable energy goals, some data centres have engineered more efficient cooling technologies. But even the hyperscale data centres that attempted to meet sustainability targets through developing advanced cooling technologies have struggled to meet expectations (Google, n.d.). As the climate warms and heat waves intensify, atmospheric control only becomes more urgent and simultaneously more difficult, which has prompted the search for environmentally sustainable data storage options beyond optimising temperature regulation.

Alternative strategies such as carbon-neutral power sources, or carbon offsets, have been adopted by big tech (Apple, Google and Meta). Companies like Meta and Amazon, for example, draw on wind farms to power their data centres. Growing energy demand of data centres is one of the most significant drivers of carbon emissions in the information and communications technology (ICT) sector globally (International Energy Agency, 2024). The share is expected to climb precipitously when we account for bitcoin mining and the current explosion in AI usage. If AI continues its current trajectory, it alone is on track to consume between 3 and 8% of the world's electricity by 2030 (Crawford, 2021; Dauvergne, 2024; Goldman Sachs, 2024; Nelson, 2023).

As the tech industry seeks innovative solutions to its mounting energy and storage challenges, emerging alternative technologies are pushing the boundaries of traditional data storage paradigms. These cutting-edge approaches represent more than incremental improvements – they signal a fundamental reimagining of how we conceptualise, create and preserve digital information. In addition to big tech data centres' own investments in ways to 'green' their way forward (Mendell et al., 2022), breakthrough resource-efficient alternative technologies ranging from 5D optical data storage (e.g., Project Silica; Microsoft, 2024), ceramics (e.g., Cerabyte, 2024) to synthetic DNA (e.g., Biomemory, Catalog DNA) are in development.

In terms of environmentally friendly data storage, synthetic DNA is among the most promising alternatives, with companies like Catalog DNA and Biomemory working to develop systems that will not only allow the use of DNA for storage but also to form the basis of low latency computing (PubNub, 2024). However, like the metaphor of 'the cloud', DNA's symbolic dimensions weigh heavily on the claims made for it. DNA figures as an aporetic cultural reference that entwines contradictory appeals to both continuity and creativity, distinctiveness and interconnectivity, the personal and the perpetual, the microscopic and mega-scaled, interior and outer spaces.

DNA's Cultural Cache(t)

The significance of biological DNA in the broadest sense is as an intergenerational carrier of information. DNA operates at a cellular level as a coded repository of a genetic system's specifications intended to persist across time. Yet, even at this very specific level, the information in DNA engages in complex ways with its environment through intricate interactions and processes. In this way, we can understand that contingencies are always intrinsic to biological inheritances.

In other words, DNA information is produced recursively and describes possible, or better, probabilistic outcomes that can't necessarily be predicted with absolute certainty.

And yet, the technologies of DNA data storage reduce this dynamic relational complexity by focusing instrumentally on the *mechanics* of DNA understood as coded information that is stored statically in molecular structures. DNA as data storage is flatly described in terms of its four bases: adenine (A), cytosine (C), guanine (G) and thymine (T), which stand in for the ones (1) and zeros (0) of binary code. In this sense, DNA as data storage alters the definition of bio-information from a relational process of dynamic transfer into a mechanistic description of DNA that is valued for its functional utility alone.

In the presentation of DNA as a data storage 'solution', a curious conceptual sleight of hand occurs in which both the container (DNA) and the contained (information or data) are both imagined as fundamentally and inflexibly immutable. We see the effect of this belief in the invariant durability of DNA when it is conjured in criminal trials depicted in movies and television. In these programmes, DNA evidence is treated as irrefutable, undeniable proof of guilt or innocence, and this mindset has spread to judges and jurors in actual courtrooms (the 'CSI effect') (Shaer, 2016). Forensic DNA experts, on the other hand, typically present their evidence in the form of statistical probabilities.

The hype surrounding DNA dates back to the Human Genome Project (HGP) in the 1990s. 'Hype' refers here to the process of building a kind of credibility around a particular subject or idea, creating a cultural consensus that this thing is to be taken seriously and giving its invocation an almost magical quality (Rajan, 2006). The completion of the HGP in the early 2000s was hailed by then US President Bill Clinton as promising an end to war, disease and strife and human suffering of all kinds (Jasanoff, 2019). Since then, DNA has been used to market a wide variety of products, from DNA testing kits, such as those produced by 23andMe and Ancestry DNA, to fitness and nutrition apps that promise to provide you with customised advice specifically tailored to your genome, and to musical compositions based on your specific DNA sequence.

DNA storage hype has extended to its data-storage capabilities. DNA storage shows promising energy efficiency since DNA can be dehydrated and stored at room temperature, contrasting sharply with conventional data centres that demand constant cooling. While a conventional data centre might use rows of hard drives or solid-state drives storing information through magnetic or electronic means at the microscale, DNA stores information at the nanoscale through actual molecular bonds. The storage density is remarkable, with theoretical capabilities of up to 215 petabytes per gram of DNA (Tung, 2017), which translates to significantly reduced physical space and material requirements compared to conventional storage methods.

The same qualities in DNA that make it so good for storing biological data make synthetic DNA a promising candidate for the storage of now-digital data. However, the scientists working on these projects almost always draw a clear line of separation between the biological form of DNA and the synthetic versions they work with. DNA – at least, biological DNA – is remarkably robust, as demonstrated by the discovery of fossil DNA (Neuman, 2021; Teague & McRae, 2024; Tenn, 2021). If synthetic DNA is similarly robust, it could potentially preserve the data we encode in it for hundreds or even thousands of years.

Behind the hype smokescreen, there is more to consider in terms of the environmental impact of DNA-based data storage. The energy costs associated with both encoding and retrieval processes in DNA-based data storage are substantial. Current DNA synthesis methods used for encoding data are energy-intensive and rely on chemical processes that carry their own environmental footprint. Information efficiency remains relatively low, with high error rates necessitating high levels of redundancy in storage (Park et al., 2023). The retrieval process requires multiple steps: rehydration, PCR amplification, DNA sequencing and digital conversion. Each of these steps consumes energy and resources, and current sequencing technologies can be particularly power-hungry (Buko et al., 2023).

As Stéphane Lemair of Biomemory told us, synthetic DNA leans on the fossil fuel infrastructure as an intrinsic part of its production:

> We [Biomemory] are not using the natural ATCG engines that are used in nature. We use heavily modified versions of these molecules that have been engineered chemically. So on nucleotides, these ATCG engines are nucleotides. There are groups, chemical groups, hydroxyls, amines, phosphates that are very reactive. So you cannot use this to make a controlled synthesis. It reacts in all directions; you cannot control it. So we are using very modified versions of these natural products. We use petrochemistry, fossil fuels, to convert them to ATCG engines that are non-natural and are modified.

There is limited research providing comprehensive environmental impact assessments of DNA-based data storage. Most studies to date focus on the potential advantages of theoretically projected storage density rather than the indirect environmental cost of practical production and implementation. Furthermore, there is as yet, no assessment of the indirect costs of DNA's remarkable capacity for efficient data compression and its ready availability. As Angela Bitting, Senior Vice President of Corporate Affairs and Chief ESG Officer at TWiST Bioscience, told us:

> You can buy bottles of phosphoramidite A, C, T, and G from different sources. These chemicals aren't difficult to procure. Other chemicals needed for creating DNA strands are also readily available off the shelf. We never had supply chain issues during COVID due to second source suppliers and large quantities, ensuring continued DNA provision for our customers.

Luccioni et al. (2025) call these indirect environmental costs 'induction effects'. Induction effects occur when increased accessibility and demand override the environmental savings made in terms of reduced water or energy consumption.

DNA storage as it is presently imagined would simply augment existing infrastructure for high-density archival solutions while complementing, rather than replacing, conventional hard drives for everyday data access and processing. Simply put, it will enable us to hoard yet more unnecessary data without meaningfully curtailing the overall infrastructural impacts of the existing system of data storage.

Solution or Stopgap?

Many of the companies investing in the creation of DNA-based storage systems – and DNA-based computing in the case of Catalog DNA or Biomemory, for example – expressed to us their desire for not only a more efficient technology but a less environmentally damaging alternative for storing and accessing the vast amounts of stockpiled data. Given the global reality of climate change, this is a laudable goal. A shift to more sustainable data storage could make a real, measurable difference in our attempts to delay the climate crisis' worst effects.

But in a context where the organisational structures, technological dimensions and commercial incentives of data storage keep adapting to reproduce and maintain an extractivist growth economy and maximise the profits of data centre owners, it is important to be circumspect. Data centres are mostly owned and operated either by large companies (cloud vendors and tech companies, real estate investment trusts or companies, governments or telcos) for their own commercial benefit or in co-location arrangements. All these companies and agencies have a direct financial interest in multiplying at scale the amount of data we store. The use of these data to grub insatiable AI models makes the financial pay-off from large-scale data storage even more flagrant. It is a case of putting company money where the AI mouth is. This manifest inducement mitigates against profound change. Instead, global investment in data retention is materialised as an everyday cultural practice such as hoarding that legitimates damaging systems, conventions and practices as 'business as usual'.

Perhaps we would do better to ask, who are the key beneficiaries from the increased data storage capacity offered by DNA technologies? Does DNA data storage alter our approach to information organisation, custodianship, governance and retention? Is the key to environmentally sustainable data storage really a matter of 'better', that is, more efficient technologies and corporate incentives, or do we need to intervene in information practices and cultures themselves?

During our fieldwork, artist Kyriaki Goni, shared similar concerns:

> I think that periodically something appears as the next solution to our big problems. This is a techno utopian and technosolutionist approach. But I have apprehensions about these promised

> 'solutions'. Scientifically speaking, I can't verify if DNA will be forever accessible or immune to obsolescence. But usually these are market-driven pitches – they're not about preservation. And I'm not sure there's evidence that DNA storage will solve the energy, water, and land use crisis caused by data centres.

In our interview with anthropologist Steven M. Gonzalez, he reflected that it is not just greed for corporate profits that is driving data storage appetites. He suggests looking at more traditional archival sentiments about information preservation from the archiving community as part of the underlying problem:

> I recently engaged with the digital preservation community, and it revealed this contrast between data storage driven by capitalist logic of redundancy for future profit harvesting, versus archivists and preservationists focused on protecting digital heritage and material culture. Bit rot occurs quickly, and this preservation needs to be done to protect cultural heritage and memory. While these approaches share practices, one is perhaps nobler. We must search for alternatives because digital data is remarkably feeble and ephemeral compared to older storage media like tape, cuneiform, or the kipu of Andean civilizations. The question becomes how to preserve our digital heritage sustainably while ensuring the data maintains integrity for future generations.

Finding a balance between preserving important cultural artefacts for future generations of humans and the kind of data hoarding we are currently engaged in on a global scale is a difficult task. In this radically interventionist information era, marked from the outset by dramatic, politically motivated government website scrubbing taking place in the United States at the time of writing in February 2025 (including the purging of all mentions of climate change), the task of harvesting deliberately disabled websites is tremendously urgent (Nietzel, 2025). Desperate retrospective archiving is the high-stakes cost of an all-or-nothing wager that bet information would outlive the environment that produced it. Within sight of this devastating gamble, it is all the more important to develop information archives that are explicitly designed and built to anticipate future information and environmental loss. Creating purposeful, environmentally sensitive ways to manage, archive and store significant data, rather than presuming all information will always be retrievable, is imperative.

Inevitably, we must contend not only with the environmental questions that digital archiving raises but also with social and cultural questions around what we preserve and who gets to decide what information is important enough to warrant preservation. It is all too easy to replicate current systems of power and oppression in these decisions – no matter how we approach the data hoarding problem, we will face not just logistical, economic and technological challenges, but sociocultural challenges as well.

Conclusion

We introduced the concept of the archival paradox to describe a central contradiction in contemporary data preservation: the more we seek to safeguard our digital present, the more we undermine the environmental and epistemic conditions that make future memory possible. This paradox is material – manifest in the carbon-intensive infrastructures of data centres and speculative storage technologies like synthetic DNA – but also symbolic, as the archive reflects social, political and economic hierarchies in determining what is preserved and what is lost. By expanding the notion of the archival paradox to include both the how and what of archiving, we foreground the urgent need for alternative frameworks that accept impermanence, foster care and reimagine digital memory beyond extraction and accumulation.

Significant attention, and associated resourcing, has been allocated to exploring and exploiting synthetic DNA as an energy-efficient solution for our current data storage needs. It remains to be seen, however, if, as an alternative, DNA-based data storage will truly disrupt the data centre industry rather than simply be absorbed by it, bringing its own environmental problems to replace those it helps mitigate.

The substantive problem of data hoarding, however, remains and has continued to grow despite more than half a century of exponential growth in computing power. Significantly less attention has been allocated to redressing this issue. Perhaps there comes a time when, rather than seeking to hold onto all data in perpetuity, we must instead learn to let some of it go. In this sense, we are echoing wider calls for IT degrowth including revised technological practices described as 'digital sobriety'. Péréa et al. (2023) pithily summarise this approach as not just using digital technologies better but also using them less.

There are alternative frameworks for thinking about how to re-imagine and re-engineer the digital archive using design concepts such as 'sunsetting' or 'graceful degradation'. For example, rather than assume that an object deemed worthy of preservation must be kept forever, we can invoke tech design principles that accept or even anticipate loss (sunsetting) or that account for infrastructural transition and decline from the beginning (graceful degradation; Goldman, 2018; Nowviskie & Porter, 2010).

In the rapidly evolving landscape of technology design, graceful degradation represents a paradigm shift from the traditional expectation of perpetual system performance. Rather than viewing system failure as an unacceptable outcome, this design philosophy acknowledges that technological infrastructures will inevitably experience decline and therefore builds in mechanisms to maintain core functionalities even as individual components or capabilities become compromised. Originating in engineering and computer science, graceful degradation suggests that a system should continue to operate, albeit at a reduced level of performance, when parts of it malfunction – much like an injured lifeform that adapts and continues to survive. This approach stands in contrast to brittle systems that completely shut down when encountering partial failures, offering instead a more resilient, adaptive model of technological design that mirrors the flexibility found in natural systems (and not unlike the dynamic relationality of biological DNA as described earlier in this chapter).

In the end, our technological quest for immortality through data preservation reveals more about our cultural anxieties than our environmental solutions. DNA-based storage promises a techno-utopian fantasy of eternal memory, yet it ultimately perpetuates the same extractive logics that have brought us to the brink of ecological collapse. The real challenge lies not in finding ever more complex ways to hoard information but in cultivating a radical humility – a willingness to accept impermanence, to recognise that not every digital trace needs preservation and that our survival may depend more on our capacity to selectively let go and to make space for new narratives beyond the insatiable archive.

Acknowledgements

This project was funded by the Social Science and Humanities Research Council of Canada.

References

Amoore, L. (2018). Cloud geographies: Computing, data, sovereignty. *Progress in Human Geography*, *42*(1), 4–24. https://doi.org/10.1177/0309132516662147

Bartlett, F., & Velkova, J. (2026). This photograph has never been in a data centre: An experimental intervention in pursuit of responsible research practice. In S.-M. Laaksonen, M. Pantti, & O. Dovbysh (Eds.), *Platforms and the planet: Digital materialities and sustainable futures*. Emerald Group Publishing Limited.

Buko, T., Tuczko, N., & Ishikawa, T. (2023). DNA data storage. *Biotech*, *12*(2). Article 44. https://doi.org/10.3390/biotech12020044

Burrell, J. (2020). On half-built assemblages: Waiting for a data center in Prineville, Oregon. *Engaging Science, Technology, and Society*, *6*, 283–305. https://doi.org/10.17351/ests2020.447

Carruth, A. (2014). The digital cloud and the micropolitics of energy. *Public Culture*, *26*(2), 339–364. https://doi.org/10.1215/08992363-2392093

Cerabyte. (2024). *Cerabyte – Store all data virtually forever*. https://www.cerabyte.com/

Crawford, K. (2021). *Atlas of AI*. Yale University Press.

Cubitt, S., Hassan, R., & Volkmer, I. (2011). Does cloud computing have a silver lining? *Media, Culture & Society*, *33*(1), 149–158. https://doi.org/10.1177/0163443710382974

Dauvergne, P. (2024). *AI in the wild*. MIT Press.

Goldman, B. M. (2018). It's not easy being green(e): Digital preservation in the age of climate change. In *Archival values: Essays in honor of mark greene society of American archivists*. https://doi.org/10.26207/d5xv-r891

Google. (n.d.). *Data centers*. https://www.google.com.au/about/datacenters/efficiency. (Accessed 12 February 2024).

Harrison Dupré, M. (2024). *Huge proportion of internet is AI-generated slime, researchers find striking*. Futurism. https://futurism.com/the-byte/internet-ai-generated-slime

Hawker, M. (2023). *A CDO call to action: Stop hoarding data—save the planet*. Forbes. https://www.forbes.com/councils/forbestechcouncil/2023/07/20/a-cdo-call-to-action-stop-hoarding-data-save-the-planet/

Hogan, M. (2016). Minus risk equals progress: The data center in the anthropocene. *Designing Media Ecology: The Anthropocene and our Post-Natural Future*, *5*(6), 46–54.

Hu, T.-H. (2015). *A prehistory of the cloud*. MIT Press.

International Energy Agency. (2024). *Electricity 2024. Analysis and forecast to 2026*. Revised version. https://www.iea.org/reports/electricity-2024

Jasanoff, S. (2019). *Can science make sense of life?* Polity Press.

Laaksonen, S.-M., & Premsankar, G. (2026). Estimating the carbon footprint of digital services. In S.-M. Laaksonen, M. Pantti, & O. Dovbysh (Eds.), *Platforms and the Planet: Big tech, digital platforms and environmental responsibility*. Emerald Publishing Limited.

Luccioni, A. S., Strubell, E., & Crawford, K. (2025). From efficiency gains to rebound effects: The problem of Jevon's paradox in AI's polarized environmental debate. *ArXiv*. https://arxiv.org/pdf/2501.16548

Mahnke, M. (2026). The eco-materiality of digital platforms. In S-M. Laaksonen, M. Pantti. & O. Dovbysh (Eds.), *Platforms and the planet : Big tech, digital platforms and environmental responsibility*. Emerald Publishing Limited.

McKinsey Consulting. (2023). *Investing in the rising data center economy*. https://www.mckinsey.com/industries/technology-media-and-telecommunications/our-insights/investing-in-the-rising-data-center-economy

Mendell, M., Hogan, M., & Verhoeven, D. (2022). Matters (and metaphors) of life and death: How DNA storage doubles back on its promise to the world. *The Canadian Geographer/Le Géographe canadien*, *66*(1), 37–47.

Microsoft. (2024). *Project silica*. https://www.microsoft.com/en-us/research/project/project-silica/. (Accessed 8 September 2024).

Mosco, V. (2015). *To the cloud: Big data in a turbulent world*. Routledge.

Nelson, F. (2023). *AI keeps using more and more energy. Where will it end?* ScienceAlert. https://www.sciencealert.com/ai-keeps-using-more-and-more-energy-where-will-it-end

Neuman, S. (2021). *Scientists say they could bring back woolly mammoths. But maybe they shouldn't*. NPR. https://www.npr.org/2021/09/14/1036884561/dna-resurrection-jurassic-park-woolly-mammoth

Nietzel, M. T. (2025). *Harvard, others saving data as Trump's team scrubs federal webpages*. Forbes. https://www.forbes.com/sites/michaeltnietzel/2025/02/09/harvard-others-saving-data-as-trumps-team-scrubs-federal-webpages/

Nowviskie, B., & Porter, D. (2010). The graceful degradation survey: Managing digital humanities projects through times of transition and decline. *Digital Humanities, 2010*. 192–193. https://dh2010.cch.kcl.ac.uk/academic-programme/abstracts/papers/pdf/ab-722.pdf

O'Brien, I. (2024). *Data center emissions probably 662% higher than big tech claims. Can it keep up the ruse?* The Guardian. https://www.theguardian.com/technology/2024/sep/15/data-center-gas-emissions-tech

Park, S.-J., Kim, S., Jeong, J., No, A., No, J.-S., & Park, H. (2023). Reducing cost in DNA-based data storage by sequence analysis-aided soft information decoding of variable-length reads. *Bioinformatics*, *9*(9). https://doi.org/10.1093/bioinformatics/btad548

Péréa, C., Gérard, J., & de Benedittis, J. (2023). Digital sobriety: From awareness of the negative impacts of IT usages to degrowth technology at work. *Technological Forecasting and Social Change*, *194*, 122670. https://doi.org/10.1016/j.techfore.2023.122670

PubNub. (2024). *What is low latency, and how to improve it*. https://www.pubnub.com/guides/whats-so-important-about-low-latency/

Rajan, K. S. (2006). *Biocapital: The constitution of postgenomic life*. Duke University Press.

Regalado, A. (2011). *Who coined 'cloud computing'*. MIT Technology Review. https://www.technologyreview.com/2011/10/31/257406/who-coined-cloud-computing/

Rilke, M. (2012). *Letters to a young poet*. Penguin.

Sachs, G. (2024). *AI is poised to drive 160% increase in data center power demand*. Goldman Sachs. https://www.goldmansachs.com/insights/articles/AI-poised-to-drive-160-increase-in-power-demand

Schön, D. A. (1983). *The reflective practitioner: How professionals think in action*. Basic Books.

Shaer, M. (2016). *The false promise of DNA testing*. The Atlantic. https://www.theatlantic.com/magazine/archive/2016/06/a-reasonable-doubt/480747/

Teague, R., & McRae, R. (2024). *Ancient DNA and neanderthals*. Smithsonian National Museum of Natural History. https://humanorigins.si.edu/evidence/genetics/ancient-dna-and-neanderthals

Tenn, C. (2021). *Paleontologists find possible dinosaur DNA*. The Scientist Magazine. https://www.the-scientist.com/paleontologists-find-possible-dinosaur-dna-69346

Tung, K. (2017). *DNA data storage landmark: Now it's 215 petabytes per gram or over 100 million movies*. ZD Net. https://www.zdnet.com/article/dna-data-storage-landmark-now-its-215-petabytes-per-gram-or-over-100-million-movies/

Uptime Institute. (2024). *Uptime institute global data center survey 2024*. https://datacenter.uptimeinstitute.com/rs/711-RIA-145/images/2024.GlobalDataCenterSurvey.Report.pdf?version=0

Chapter 6

This Photograph Has Never Been in a Data Centre: An Experimental Intervention in Pursuit of Responsible Research Practice

Flora Mary Bartlett and Julia Velkova

Linköping University, Sweden

Abstract

Digital platforms have become an elemental part of life – distributing media content, organising sociality, undergirding energy, transport, urban development and food distribution. Their operations today are materially anchored in landscapes, communities and electric grids through the data centres that store the computers upon which platforms run. It is well known that data centres are entangled in extractive resource practices. Yet so is critical scholarly practice, contributing to conflicts about landscape, labour and energy. In this chapter, we combine methods and insights from visual anthropology and media studies in an interdisciplinary intervention, presenting an experimental approach to photographing data centres in Sweden in which we endeavour to avoid image storage that contributes to the centres' troubled environmental relations. We experiment with analogue photography as a method, addressing the possibilities for researching digital materialities without using online platforms, the materials needed, the epistemic potential and the ongoing challenges of working with non-digitised images in the publication process. We ask what could be considered a responsible visual methodology in this field, and how we can balance the different material demands of photographic practice. We explore if it is possible to be critically engaged when researching digital platforms and infrastructures without contributing to the conflicts that these online platform material networks ignite, what this means and the implications for research practice.

Platforms and the Planet, 99–113

doi:10.1108/978-1-83662-172-020261007

Keywords: Sustainable digital research practice; visual methodology; data centres; media studies; visual anthropology; digital infrastructure

Introduction

> One rainy morning in September 2023, we took a drive to Amazon's data centre in Katrineholm – an unremarkable sight in an equally unremarkable Swedish town, known primarily as a railway hub connecting the east and west coast of the country. We intended to observe the Amazon site and its surroundings in order to understand the scale of its land and energy usage in the region. No signs revealed its ownership, generally unsurprising in the field of big tech, but the sign that did meet us was unambiguous: 'NO PHOTOGRAPHY'. Several US-prison style mounted security cameras surveilled the area and, presumably, us. We felt the tendrils of aggression snaking from these infrastructural accessories as we simultaneously became subjects of the all-seeing vision of the centre, yet censored in our ability to return the gaze.
>
> This was an immediate provocation for two researchers working variously with data infrastructures, environmental relations, and visual methods. How could we study this place and engage with it visually when we were so clearly unwelcome observers? Disregarding the warning, we snuck to higher ground from which we could photograph the building. Flora contemplated the consequences of shooting analogue – using an old Nikon from the 1990s loaded with a roll of Fujifilm as part of her practice as a visual anthropologist. She has found analogue very practical, lightweight, requiring little battery power, and having high-quality printing capabilities with complex colour palettes but without the need for an expensive digital camera. During this excursion, analogue photography also offered an unexpected moment of reflection. Wound up inside the Nikon was a roll of exposed negatives that, once developed, would show the vast façade of the data centre and the electricity transformers and contentious powerlines through which this facility has come to occupy ¼ of the grid capacity of the town, invisibly steering not only who can see and know of Amazon in the town but also which industries could or could not be part of the town's future. But at that precise moment in time, the roll was still a non-digital, physical trace. If it was never scanned and saved onto a server, then it would not traverse through a data centre, nor would it occupy space on Amazon's, or someone else's, infrastructural platforms. But what would it mean to claim that this photograph has never been inside a data centre?

Online platforms organise and influence many aspects of life: from everyday mobility, communication and entertainment to the logistics of global commerce and electricity provision. Yet platforms are also materially anchored in landscapes, communities and electric grids through data centres such as the Amazon Web Services (AWS) data centre in Katrineholm, which stores the computers upon which platforms run and provide web services to programmers all over the world.

The centre's operations are sustained through the corporate cultivation of water, earth and air in the service of machines that produce and operate online services 24/7. A decade of research into the cultural and material politics of data centre construction, including our own, has explored these troubled relations of platform infrastructure as they relate to place and environment, and how their production is entangled with the making of both desirable and threatening futures for humans and more-than-humans (Hogan, 2018; Johnson, 2023; Lehuedé, 2022; Velkova, 2021). Scholars have extensively documented how groundwater aquifers, municipal water infrastructure, fisheries and oceans are routed to refrigerate fragile computers placed in data centres (Gilmore & Troutman, 2020; Hogan, 2018; Sutherland & Bopp, 2023). In northern regions like Sweden, the force of rivers, captured in the form of hydropower electricity production facilities, is mobilised to supply computers in centres like AWS with electricity earmarked as 'green' and thus presumed to be 'good', while building on legacies of Sámi people's land dispossession and salmon expulsion. In other parts of the world, coal and oil – the fossilised temporalities of the past life on earth – are burnt at ever-growing rates to generate the electricity for servers that enable someone's meal to be ordered online or someone else to share a selfie. These infrastructural and material processes sustain digital platforms through an ongoing push on people, animals, plants and industries who are reorganised and displaced in order to leave way to the carnivorous demands of computation infrastructure in the name of convenience and business that platforms offer. If we are to question the environmental responsibility of digital platforms, it is not enough to diagnose these conditions as scholars. We must also question the materialities of our own scholarly practice and their inextricable embeddedness in and co-production of industrial-scale data processing.

Much of the research on internet infrastructure and data centre industries, including our own, remains reliant on those same infrastructures. Critical scholars contribute to those very same conflicts around land, water, air and energy through the materialities and supply chains of producing scholarship about these practices.[1] A photo of a data centre taken with a mobile phone ends up in another data centre almost instantaneously. When a critical scholarly

[1]While we write from the positionality of researchers in the Global North, this point is relevant for scholars around the world utilising the cloud as part of their research practice. Our choice to study the realities of this in Sweden is motivated in part by the data centres' focus on being sustainable in this context, and yet as we show here, they are entangled in problematic relations and supply chains across the globe. This is also reflected in our choice of works cited that pick up this global scale of data relations.

article about the data centre gets published online, it is reproduced via hosting facilities that ultimately lead to either a Google, AWS or Microsoft-owned data centre. The cultures of producing and preserving scholarly knowledge, its materials and data are all part of the problem of generating demand for more data centres (Hogan, 2021), thus more electricity, more water, more space, more of everything, leaving less breathing space (Savolainen & Ruckenstein, 2022) for other industries, cultures and ways of life. This is not only a problem in research but also for digitally mediated life in general that has been profoundly reorganised in the past 15 years towards remote server storage and computation (Parks et al., 2023).

In this chapter, we experiment with analogue photography as a method to provoke critical reflection upon the possibilities for doing research on digital materialities without engaging online platforms and data centres. The question of methods through which digital infrastructures and platform relations come to be known has been pertinent, particularly when it comes to research on data centres. Access to data centres is most often restricted, limiting the scholars' possibilities to talk to the companies or do fieldwork at these sites veiled in secrecy (although exceptions exist, see Gonzalez Monserrate, 2023; Taylor, 2022; Velkova, 2020). Therefore, scholars have argued for finding other routes, exploring relations and cultures of data 'peripheries': the industrial mobilities, 'tentacular relations' (Bartlett, 2024), supply chains and labour that feeds into data centres, including the rare minerals and hard drive economies that make data (Cooper, 2021).

We expand this methodological discussion through a call for more interventionist methods that are sensitive to how our own scholarly practice activates digital materialities and intertwines with them. We take the photo roll as an evocative object (Turkle, 2007) to think with and about the limited possibilities of research to disconnect critical scholarly discourse from the contentious politics of digital infrastructure in which it is embedded. We share the dilemmas and reflections that arose from our attempt to use the photo roll and analogue photography in research on data infrastructure. Furthermore, we use our analogue photographs of the AWS data centre to reflect on the materials and webs of relations needed for scholarly practice, the ongoing challenges of working with non-digitised or non-data-centred images in the publication process and the costs of cultivating environmental response-abilities in research on digital infrastructure. Beyond the environmental impacts of photography production, this experimental intervention speaks to a larger question of sustainable practice among our research group. Can we be critically engaged researchers without contributing to the material networks of online platforms? What are the implications of such a provocation? Our intent here is not to give definitive answers or prescriptions but rather provoke a conversation about the webs of material-environmental relations that different methodological choices invite and the possibilities and costs of cultivating environmental response-abilities in research on digital infrastructure.

Documenting: Digital vs Analogue Image-Making in Research

Visual anthropology calls for multimodal scholarship, combining different media such as sound, image or performance in research practice to 'generate relations' and lead to new avenues of inquiry (Dattatreyan & Marrero-Guillamón, 2019, p. 220). Beyond simply illustrating written texts, multimodality can contribute with something different. It can become a 'tactic of transmission' (Jungnickel, 2020, p. 3), where transmission is understood as both the machinery for research in engaging and exploring the troubled energy relations of data infrastructure, and for communicating the knowledge created in the process. A key part of our use of this methodology involves an experimental approach that offers a 'distinctive ethnographic modality' (Sánchez Criado & Estalella, 2018, p. 13), which gives new insights into environmental relations (Bartlett, 2021).

In our research, we relied on photography as part of a multimodal inquiry into the relations of AWS to the place where it stands today, Katrineholm. Photography demands consideration of the *afterlife* of the images we produce: how they are mediated after their production by other institutions 'fraught with ethical dilemmas' (Collins et al., 2017, p. 142) such as universities, galleries or museums. Once images inhabit other contexts, they take on new characteristics or are framed in new ways. We should expand this understanding and critically address the afterlife of images that we generate or use in research, both digitalised and analogue – how they exist in the world after their production and what environmental afterlives they leave in their wake. The purpose of our intervention is less about avoiding digitalisation per se and more about actively engaging materialities that do not as easily *lend themselves* to digital platforms. A digital image is stored on a memory chip in megabytes before being transferred to a personal computer and potentially the cloud. An analogue image is transferred temporarily from a strip of film to a local studio computer to print out the images (unless one makes direct darkroom prints). What difference does it make?

Digital photography's main ecological problem today is the rate at which the world takes photographs and the storage requirements this creates. Following the JPEG revolution in 1992, image storage capabilities mean we now make 1.6 trillion new photos per year, 90% of which are taken on smartphones and end up in personal storage or on social media (Sneyers, 2023). The material demands of digital visual imagery are substantial. Digital photography has higher demands on manufacturing and, notably, storage than analogue. At the manufacturing stage, digital cameras require more electric components than analogue, and which involve toxic substances and precious metals such as lead (Urth Magazine HQ, 2022). These cameras are upgraded quickly, with an average lifespan of five years (Sneyers, 2023), leading to landfill for both cameras and batteries and causing chemical contamination that will last decades (Gabrys, 2013). While smartphones increasingly replace digital cameras, reducing the need for more equipment for the average user, the ability to make large prints from these images is still limited for those making images for art or research. In contrast, Flora's analogue camera has been in use since the early 1990s and requires a battery change every two to three years, and the small negatives can make large prints using darkroom or digital technologies of the day without relying on pixels that risk becoming obsolete.

Digital image storage also requires the material infrastructures of data centres, far from having a remote, ephemeral existence in a dematerialised 'cloud'. With the average mobile phone image requiring around four megabytes of storage, the 1.6 trillion images equate to 10 exabytes of storage, which requires 20 terawatt hours of energy (20 million Megawatt hours) just to store the images in the cloud, before even considering the added storage needs for image sharing (Sneyers, 2023), or the computing capacities that trending image-generating artificial intelligence (AI) requires. This energy needs to come from somewhere – a hydropower plant on reindeer herding land in Northern Sweden or an Amazon solar power plant placed on sacred land in South Africa (Ortar et al., 2023) before it is delivered via a power grid to a data centre placed in a town like Katrineholm, for example. By proliferating digital images of these infrastructures, we perpetuate such relations connecting data centres elsewhere. The 'afterlife' of digital images is thus as data requiring energy, contributing to new energy infrastructures, local conflicts and grid competition (Libertson et al., 2021). For example, unlike the ease of generating a digital image and storing it in a camera and then a 'cloud', electricity grids have capacity limits. Those in Katrineholm are today partitioned: ¼ goes to Amazon and another sizeable chunk to its neighbour by powerline, Kronfågel, known as the largest chicken slaughterhouse in Sweden, that kills and grills 200,000 chicken per day. The remaining grid capacity is to be split between all other existing and new businesses, making Amazon and the chicken business regulate (via their stakes in the grid) who can come to town and who cannot.

Energy is also needed to produce the resources and manage the legacies of building the servers, including lithium, platinum, aluminium, silicon, copper, boron, cobalt, tungsten and tantalum (Parikka, 2015). These materials are entangled in their own violent politics of human and environmental catastrophe elsewhere. Rather than the popular imaginary of a fluffy, green and clean cloud (see Wyatt, 2021), Sy Taffel writes of the 'miasma of toxic smog' caused by the dependence of cloud computing on oil to extract and purify these materials (Taffel, 2021, p. 987). Each digital image being taken activates these networks of energetic and socioeconomic relations that produce, display and circulate an image.

This is not to say that analogue imagery is a white knight. Film processing has its own material legacy and environmental afterlife, such as the plastic film casing, the chemical makeup of the negatives and the fixer and developer required to process the film in the lab. It is, as Sneyers writes, 'not an eco-friendly technology' (2023). Many of the chemicals are toxic, notably the silver ions used on the negatives, and these chemicals are often washed into the drain during processing (Urth Magazine HQ, 2022). These chemical afterlives then continue in the water systems. Notably, Stockholm's Water Authority measured a 50% decrease in silver pollution, arguing the popularity of digital photography as a major factor (Urth Magazine HQ, 2022). Artist and anthropologist Alice Cazenave is exploring the chemical demands and impacts of analogue processes through her work with The Sustainable Darkroom and her PhD research[2] exploring the social histories of settler colonialism in the case of Kodak film production in the United States. Such work sheds light on the problematic practices and sociomaterial footprints of this method, complicating its place as a romanticised 'return' to something purer and more eco-friendly.

Having discussed the relative ecological costs of both these technologies, what are the ways forward for *producing* research images while also considering their afterlives? There are ways to take the afterlives of these methods into account and have more responsible approaches. In digital photography, manufacturers are taking steps to phase out toxic chemicals, cameras can be disposed of in less harmful ways, and rechargeable batteries can be used (Urth Magazine HQ, 2022). With analogue processing, less toxic chemicals are being explored, including plant-based developers as demonstrated by The Sustainable Darkroom using pine and algae.

Applicable to both methods, and key to this chapter, is the way in which we approach image taking itself. We want to argue for a scholarly practice informed by material limits, where we recognise that the production of research material has real-world material consequences, in contrast to the common imaginary of limitless and costless data generation perpetuated by cloud services. As scholars, we should be conscious in our choices to take *fewer images* when using digital photography and not upload everything to the cloud by default. Returning to the idea of technologies that do not as easily lend themselves to digital storage, analogue methods aid this choice. Film is costly and rolls are limited to 24 or 36 exposures. This is a method in which mindful usage feels more tangible, and taking more than one photograph of a scene or event is hindered by the limitations of the roll. In the 2018 Swedish comedy *Tårtgeneralen* ('The Cake General') set in 1985, the local photographer consistently covers village events with one single click of the shutter followed by the brisk words '*Tack, det räcker*' – *Thanks, that's enough.* He then swiftly exits the scene to the surprise of the locals expecting several shots. This is, perhaps, a very pertinent motto to adopt when taking photographs, as a reminder to critically engage with our image afterlives and production and to realise that yes, thanks, one is in fact enough.

Embedded in the Cloud

Modern academic practice is increasingly cloud-based, with file sharing, virtual and hybrid-meeting formats and large volumes of email becoming the norm. Even without visual media, the data centre is already utilised in our research practice. In our team, we use Microsoft services, including Teams and OneDrive, as well as Zoom and an external website. We email abstracts to one another and save relevant material in a shared server. We started to ask ourselves where

[2]The Sustainable Darkroom is an artist-run programme working towards environmentally friendly photographic darkroom practice through workshops, talks and training. See www.sustainabledarkroom.com and www.alice.cazenave.co.uk.

exactly our files are stored, and if it is possible to 'visit' them – to see them occupying space. In asking these questions, an uncomfortable truth emerged around the hybrid-meeting table. We did not have the power to visit our data, to see where it was stored with our own eyes or even to find out exactly where in the world our files spent their time. Our research project on data centres' energy use and materialities was itself deeply entangled in the culture of opacity of this industry, images or no.

To avoid becoming complicit in it, we would have to refrain from sharing files unless via a hard drive or sending printed copies to one another. Avoiding these services would require resource use in another way, like increased paper use and the hardware for storing and exchanging files locally. It would also be less practical in processes of co-authoring or sharing information quickly. It would also mean solely in-person meetings, obviously problematic given the carbon cost of academic travel (see Nevins et al., 2022). It would transform our practices of doing and communicating research all together. And perhaps it should.

The Digital Humanities Climate Coalition (DHCC) is a group of UK scholars who are critically addressing the environmental implications of digital humanities research and have produced a toolkit for mitigating academic impacts on the climate (Baillot et al., 2024). In this collaborative document, the authors present suggestions for cutting down on carbon across multiple research activities 'while also critically reflecting on their nuances, in the broader context of climate justice'. While acknowledging the carbon emissions of videoconferencing, they highlight how travel emits even more. Making hybrid or online meetings more environmentally friendly, they argue, can be achieved by turning off cameras for parts of the meeting or choosing a more ethical videoconferencing host. The goal is balance and reflectiveness rather than an all-or-nothing approach. They also argue that sharing files via the cloud is more carbon-efficient than emailing copies back-and-forth, as an email weighs four grams of CO_2, up to 50 grams with a photo attachment.

This approach is specific to the calculable costs of *carbon* in research. What our intervention plays with, in contrast, is a specific reaction to the place-based environmental impacts of data infrastructure. This is not limited to carbon but addresses a multitude of other assemblages as discussed above – materials, land use, energy infrastructures and local conflict. A tricky situation thus ensues: by selectively avoiding the cloud, we avoid certain emplaced environmental impacts and energy usage, but by avoiding it completely, we cause higher carbon costs and environmental destruction elsewhere through the necessity of more travel and increased printing. Perhaps the resolution lies not in avoiding the cloud completely, but – as the DHCC advocate – being critical and reflective in our practices instead of falling for 'techno-utopian, carbon-intensive technological traps' (Baillot et al., 2024).

While the DHCC's recommendations are commendable and thorough, and individual researchers should indeed be critical of their practice, we must also keep in mind the structures that facilitate and encourage resource-intensive practice. As environmental damage is a systemic problem, we must also look beyond the level of the individual in addressing *both* the climate crisis and the problematic assemblages of the cloud. University-wide structures and IT departments that physically situate research in cloud-based services are also complicit and must question which digital services they employ, and the size of the capacities in data centres they demand. It is not enough in our time of environmental collapse that choices are made based on economic value or production speed alone. The use of analogue images to explore platform materialities like data centres exposes these webs of relations and difficult questions for own research practices. And, as with the visual imagery, the answers to these questions are complicated. Rather than trying to avoid the data centre assemblages completely, we must critically assess our dependence on them. Following the DHCC, there are measures to reduce our data usage as well as our carbon emissions (see also Vigren et al., *this volume*), but we must also critically engage with the larger forces at work: land use, energy intensity, grid competition, fossil fuel-run backup generators and resource extraction (see also Premsankar & Laaksonen, *this volume*). As scholars of media and environmental anthropology, we also have an academic responsibility to engage with these topics and make visible these problematic processes as well as address how our own practice contributes to – or interrupts – them.

Where article sharing and emails are difficult to avoid, storing and sharing images feels like one achievable intervention in this complex web of cloud-based research practice. It can thus serve as one such modification to an otherwise cloud-heavy culture of academia.

Exploring

What does analogue photography do to the ways we know data infrastructure? We workshopped this question during a project meeting of our research group, surrounding a table upon which Flora placed the bundle of her developed photos from our visit to the Amazon data centre and its surroundings. It was a thick stack that quickly got scattered and messy. We carefully took a photo and examined it, holding it in our hands, careful not to leave fingerprints on its surface. We asked each other *which is your favourite image? What do you see?* The photographs were very different from the commercial imagery that is circulated online. The dull but nuanced shades of grey and black, of the sky and the façade of Amazon's data centre printed on photographic paper, gave us a sense of authenticity and the feeling of a 'better', more 'truthful' representation of our experience/encounter with Amazon in Katrineholm.

There is no inherent or objective 'truth' to an image, nor can it be removed from its production and the social entanglements at play in the choices of the photographer (Tagg, 1993). A scene is interpreted by the person behind the camera, and then translated into 2D by the mechanisms of the camera itself (Tagg, 1993). Flora chose the type of film to use, where to stand, how to frame the photograph and to develop them as matte prints with white borders. The photographs were produced through existing relations with photography, place, Amazon, the weather and the lab technician who developed them. In other words, through the sociocultural entanglements of the images in what lies 'behind' the paper: the 'complex fabric of notions, representations, images, attitudes, gestures and modes of action' regarding the images' production (Tagg, 1993, p. 100). Likewise, photographs are always received within a particular cultural frame, sparking associations or values particular to that context. We must therefore consider the material-discursive systems photographs are a part of and the institutions with which they intersect. In this case, they were received within the specific context of the research group's meeting, building on existing conceptions of analogue photography. Care and awe surrounding the analogue images may well have stemmed from several cultural associations among the group, including a notion of scarcity in contrast to the plethora of digital images in our world, but also a mysticism of the analogue process itself and its blurred position between the categories of art and documentary.

We also spent time with each photograph, discussing and showing it to each other, passing it from hand to hand, a ritual that would have taken a different social and temporal dimension and relation to the image had we done it with a scroll of online images. This mode of engaging with our research 'data' resembled what media scholar Laura Marks (cited in Parks, 2007) calls haptic visuality:

> As Marks (2002) explains, 'Haptic visuality, a term contrasted to optical visuality, draws from other forms of sense experience, primarily touch and kinesthetics' (p. 2). She invokes the haptic to critique post-Enlightenment rationality and its privileging of distant and detached modes of observation and sets out to 'restore a flow between the haptic and the optical that our culture is currently lacking. . . [and] to explore how a haptic approach might rematerialize our objects of perception' (p. xiii).

Our use of analogue photography enabled us to touch 'the cloud' sensorially and visually, and pass it around, while remaining at least for a moment infrastructurally detached from it.

Exploring the images together in the research group in this way opened new avenues of discussion and reflection surrounding the knowledge and representations of data centres. By handling the physical prints, we were inspired to discuss the materialities consisting of, within and beyond the data centre represented in the prints, as opposed to the dominant metaphor of the abstract and ephemeral cloud. We reflected over the gritty industrial reality of this space, in stark contrast to the glossy digital images circulating from within the industry. The meeting room table became an 'arena of exchange' (Bourriard, 2002), mirroring that of an exhibition where a space is created between and around the images in which unexpected conversations are generated. Such artistically informed and multimodal methods thus create new inroads and ways of thinking around the subjects of our research, forcing us to reflect on our engagement with these infrastructures and the implications of resisting them.

Challenges of Cloudless Multimodality

Presenting the analogue images from this intervention was no simple task, and in navigating the path of cloudless multimodality, we stumbled upon unexpected complications that led to unforeseen and fruitful discussion.

Originally, the idea was as simple as not digitising the lab prints or uploading them to our shared server. They existed in two forms: the negatives processed and stored in a plastic sheet and single copy prints made by the lab with no digital copies kept. When intending to show the images to the team, we suddenly realised there were further limitations. Everyone would have to be in person at Linköping University to see the images, otherwise they would end up being shown during a Teams meeting hosted by Microsoft and thus in a data centre – though, interestingly, not necessarily in the same country as the meeting (Arbuthnot, 2019).

When considering this very chapter, we realised we could not show the photograph here. The publishing process utilises data sharing and storage through the necessary back-and-forth of emails providing valuable feedback on drafts, peer-review and layouts. The final version would be saved on a server for printing and open access – occupying a small place in a data centre somewhere in the world. To fully invest in this experiment, you the reader may not view this image either (see Fig. 6.1), unless you happen upon Linköping University campus to view the little package of prints safely tucked away in an office or stumble upon us presenting this chapter at an in-person conference.

Moving these images out of the sphere of research and into an artistic environment such as the gallery or museum would be a way to experiment with this

Due to Ethical and Methodological Concerns,
This Photograph was Never Here

Fig. 6.1. As Displaying the Photograph Here Would Immediately Secure its Place Within a Data Centre, we Used This Simple Figure for the Duration of the Writing and Publishing Process.

intervention in a larger discursive exhibition space. With such an exhibition, we would respond to Hogan's (2021) call to engage art in the study of data centres on a larger scale, reaching a larger public without entering the infrastructures of the data centre itself. However, with such an exhibition, we would invite the risk of viewers taking photographs of these images and saving them to the cloud via social media or their own personal storage. The photographs would thus end up in a data centre, changing this experiment. This conundrum raises questions about what a digitised image is. Fiona Cameron (2019, p. 66) has argued that digitisation does not produce a copy but rather a new 'thing'. If a digital scan of an analogue photograph is an altogether new image, then digitising these analogue images of the data centre would not have resulted in *the same image* ending up in a data centre. The result, however, would still be *an image* stored in the cloud and would thus end (or fundamentally change) this intervention. We must perhaps take our chances, using signage to urge viewers not to take photographs in the hope that they view it as a chance to join our small intervention and allow this experiment to continue.

Conclusion: Implications for Reflexive, Responsible Research Practice

One of the many benefits of multimodality is to bring out the unforeseen in research practice and lead us down paths we did not expect. This is true of our excursion to the AWS data centre: the analogue camera in relation to the aggressive messaging provoked an unexpected reflection, resulting in discussions that later became this very chapter. Despite Susan Sontag's (1979) famous assertion that photography is by default a non-intervention, here it served as a methodological interference into our daily practices as researchers.

What implications does this interference have for future research, and acknowledging how we occupy these problematic material-environmental webs of relations? As the authors of the DHCC document make clear, there is 'not one right answer' for such conundrums (Baillot et al., 2024). Recognising then the challenges in avoiding the cloud completely, as well as the demands of images in cloud storage, we consider the following practice. We will continue to use OneDrive for saving documents, for emailing, for hybrid meetings, for publication and for file storage as that is the cultural and material context in which we academics currently reside. But in the same way as we take care in considering the ethics of the research we do, we will try to think before we store, before we backup, before we duplicate and share folders filled with images and copies. In the case of these particular images, it equates to roughly 180 megabytes that will resist the cloud.

In their examination of the anatomy of AI, Crawford and Joler (2019) acknowledge that accounting for resource costs is impossible, but they assert that it is 'increasingly important that we grasp the scale and scope if we are to understand and govern the technical infrastructures that thread through our lives' (Section II). While we can tangle ourselves in knots making the calculations of the environmental costs of our research, the crucial thing is to recognise these infrastructural threads and how our research is interwoven with them if we are to remain critical towards these industries.

Our aim here is not to discourage the use of visual material in research. Far from it in fact, when we know how multimodality can expand our research methods both in the field and in disseminating critical perspectives to diverse publics. Instead, we urge scholars to be mindful of how much they create, store, share and backup. Is all of it necessary? For how long? Ultimately, we wish to provoke a reflexive scholarly practice that challenges the notion of limitlessness so often presented by tech companies in their idea of an ever-expanding, weightless cloud of data.

References

Arbuthnot, T. (2019). *Where in the world will my Microsoft Teams meeting be hosted?*. Tom Talks. https://tomtalks.blog/where-in-the-world-will-my-microsoft-teams-meeting-by-hosted/

Baillot, A., Baker, J., Bunn, J., Cline, A., Faerber, M., Feidicker, C., Lethbridge, J.; McConkey, M.; Moore, J.; Olge, C., Roeder, T., Seymour-Smith, N., Steer, M., Walton, L., & Williamson, E. (2024). Introduction. *The Digital Humanities Climate Coalition toolkit for Climate Justice in humanities research.* https://sas-dhrh.github.io/dhcc-toolkit/

Bartlett, F. M. (2021). In galleries and beer cans: Experimental photography in explorations of Northern Swedish landscapes. *Visual Anthropology Review*, *37*(1), 164–192. https://doi.org/10.1111/var.12236

Bartlett, F. M. (2024). *The tentacular data centre: Tracing the global and local materialities of the cloud.* Presented at the Antroforum Seminar, Department of Thematic Studies, Linköping.

Bourriard, N. (2002). *Relational aesthetics*. Les presses du rèell.

Cameron, F. R. (2019). Theorising heritage collection digitisations in global computational infrastructures. In Lewi H., Smith W., vom Lehn D., Cooke S. (Eds.), *The routledge international handbook of new digital practices in galleries, libraries, archives, museums and heritage sites*. Routledge.

Collins, S. G., Durington, M., & Gill, H. (2017). Multimodality: An invitation. *American Anthropologist*, *119*(1), 142–146. https://doi.org/10.1111/aman.12826

Cooper, Z. G. T. (2021). Of dog kennels, magnets, and hard drives: Dealing with big data peripheries. *Big Data & Society*, *8*(2). https://doi.org/10.1177/20539517211015430

Crawford, K., & Joler, V. (2019). Anatomy of an AI system. *Virtual Creativity*, *9*(1), 117–120. https://doi.org/10.1386/vcr_00008_7

Dattatreyan, E. G., & Marrero-Guillamón, I. (2019). Introduction: Multimodal anthropology and the politics of invention. *American Anthropologist*, *121*(1), 220–228.

Gabrys, J. (2013). *Digital rubbish: A natural history of electronics*. University of Michigan Press.

Gilmore, J. N., & Troutman, B. (2020). Articulating infrastructure to water: Agri-culture and Google's South Carolina data center. *International Journal of Cultural Studies*, *23*(6), 916–931. https://doi.org/10.1177/1367877920913044

Gonzalez Monserrate, S. (2023). Thermotemporalities, thermomasculinities: Uptime, downtime, and server heat in the digital Anthropocene. *New Media & Society*, *25*(2), 324–344. https://doi.org/10.1177/14614448221149938

Hogan, M. (2018). Big data ecologies. *Ephemera*, *18*(3), 631–657.

Hogan, M. (2021). The data center industrial complex. In M. Jue & R. Ruiz (Eds.), *Saturation: An elemental politics. Elements* (pp. 283–305). Duke University Press.

Johnson, A. (2023). *Where cloud is ground: Placing data and making place in Iceland*. University of California Press.

Jungnickel, K. (Ed.). (2020). *Transmissions: Critical tactics for making and communicating research*. The MIT Press.

Lehuedé, S. (2022). Territories of data: Ontological divergences in the growth of data infrastructure. *Tapuya: Latin American Science, Technology and Society*, *5*(1). https://doi.org/10.1177/14614448221149938

Libertson, F., Velkova, J., & Palm, J. (2021). Data-center infrastructure and energy gentrification: Perspectives from Sweden. *Sustainability: Science, Practice and Policy*, *17*(1), 153–162. https://doi.org/10.1080/15487733.2021.1901428

Nevins, J., Allen, S., & Watson, M. (2022). A path to decolonization? Reducing air travel and resource consumption in higher education. *Travel Behaviour and Society*, *26*, 231–239. https://doi.org/10.1016/j.tbs.2021.09.012

Ortar, N., Taylor, A. R. E., Velkova, J., Brodie, P., Johnson, A., Marquet, C., Cirolia, L. (2023). Powering 'smart' futures: Data centres and the energy politics of digitalisation. In S. Abram, K. Waltorp, N. Ortar, & S. Pink (Eds.), *Energy futures: Anthropocene challenges, emerging technologies and everyday life*. De Gruyter.

Parikka, J. (2015). *A geology of media*. University of Minnesota Press.

Parks, L. (2007). Points of departure: The culture of US airport screening. *Journal of Visual Culture*, *6*(2), 183–200. https://doi.org/10.1177/1470412907078559

Parks, L., Velkova, J., & De Ridder, S. (Eds.). (2023). *Media backends: Digital infrastructures and sociotechnical relations*. University of Illinois Press.

Sánchez Criado, T., & Estalella, A. (2018). Introduction: Experimental collaborations. In T. Sánchez Criado & A. Estalella (Eds.), *Experimental collaborations: Ethnography through fieldwork devices* (pp. 1–30). Berghahn Books.

Savolainen, L., & Ruckenstein, M. (2022). Dimensions of autonomy in human–algorithm relations. *New Media & Society*, *26*(6). https://doi.org/10.1177/14614448221100802

Sneyers, J. (2023). History and environmental impact of digital image formats. *The Photographers'g Gallery: Unthinking photography*. https://unthinking.photography/articles/history-and-environmental-impact-of-digital-image-formats

Sontag, S. (1979). *On photography*. Penguin.

Sutherland, T., & Bopp, G.. (2023). The Pacific futures of subsea data centers. *New Media & Society*, *25*(2), 345–360. https://doi.org/10.1177/14614448221149944

Taffel, S. (2021). Data and oil: Metaphor, materiality and metabolic rifts. *New Media & Society*, *25*(5), 980–998. https://doi.org/10.1177/14614448211017887

Tagg, J. (1993). *The burden of representation: Essays of photographies and histories*. University of Minnesota Press.

Taylor, A. R. E. (2022). Cloudwork: Data centre labour and the maintenance of media infrastructure. In E. Costa, P. G. Lange, N. Haynes, & J. Sinanan (Eds.),

The Routledge companion to media anthropology (pp. 213–228). Routledge. https://doi.org/10.4324/9781003175605-21

Turkle, S. (2007). *Evocative objects: Things we think with.* MIT Press.

Urth Magazine HQ. (2022). Is Analogue or digital photography more eco friendly? *Urth Magazine.* https://magazine.urth.co/articles/analogue-versus-digital-photography-eco-friendly

Velkova, J. (2020). The art of guarding the Russian cloud: Infrastructural labour in a Yandex data centre in Finland. In *Digital icons: Studies in Russian, Eurasian and Central European new media* (Vol. 20). https://www.digitalicons.org/issue20/the-art-of-guarding-the-russian-cloud

Velkova, J. (2021). Thermopolitics of data: Cloud infrastructures and energy futures. *Cultural Studies*, *35*(4–5), 663–683. https://doi.org/10.1080/09502386.2021.1895243

Wyatt, S. (2021). Metaphors in critical internet and digital media studies. *New Media & Society*, *23*(2), 406–416. https://doi.org/10.1177/1461444820929324

Chapter 7

Platform Opacity in the Global E-waste Dilemma

Lauren E. Bridges[a] *and Vusumuzi Maphosa*[b]

[a]University of Virginia, USA
[b]National University of Science and Technology, Zimbabwe

Abstract

The increasing platformisation of retail has profound implications for the global e-waste crisis, which is projected to reach 82 million tons by 2030. This chapter investigates the role of retailer-platforms in the dynamics of digital device production, consumption and disposal, emphasising their influence on e-waste management practices. Through corporate reports and semi-structured interviews with retailers in the United States and Zimbabwe, the authors reveal contrasting approaches to e-waste in the Global North and Global South. In the United States, platformised retail environments often obscure recycling initiatives, shifting responsibility to state governments with limited infrastructure. In contrast, Zimbabwean retailers primarily operate in offline markets, where inadequate policies and recycling systems place the burden of e-waste management on consumers and informal networks. The analysis highlights how the platformisation of retail not only accelerates consumption but also exacerbates global inequities in e-waste accumulation and disposal. The study underscores the opacity of platforms in managing e-waste, which creates ambiguity around how recycling programmes function and allows platforms to evade responsibility for their role in e-waste management. To address this platform opacity, the authors propose an expanded retailer platform responsibility to tackle information asymmetries and enhance accountability within the tech supply chain. By focusing on the often-overlooked role of retailers, this chapter contributes to a more comprehensive understanding of the sociotechnical politics surrounding e-waste and advocates for a collaborative approach to mitigate its impacts.

Platforms and the Planet, 115–131

doi:10.1108/978-1-83662-172-020261008

Keywords: E-waste; recycling; platforms; platformisation; retail; digital marketplaces; extended producer responsibility; environmental sustainability; Global South

Introduction

The infrastructural turn in digital platform studies has drawn attention to the materiality of the internet and the sociotechnical politics that sustain digital infrastructures (Plantin et al., 2018). Within this work, scholars have unpacked the digital wastes entangled with the production, consumption and discarding of digital devices (Gabrys, 2013; Lepawsky, 2018). Other scholars have highlighted global inequities in the flow of e-waste from the Global North to the Global South (Baldé et al., 2022), inequities in inter-regional flows (Lepawsky, 2015) and harmful local discarding practices (Koebler, 2016). As the largest growing waste stream on the planet, e-waste is estimated to reach 82 million tons (MT) by 2030 (Baldé et al., 2024). E-waste contains valuable and reclaimable materials, such as precious and rare earth minerals and plastics, and hazardous materials, such as bromine, cadmium, chromium and arsenic, which present health risks to communities and local ecologies if not disposed of properly (Maphosa & Mashau, 2023).

Research on e-waste interventions has primarily concentrated on the responsibility of producers regarding the end-of-life disposal of their products through the extended producer responsibility (EPR) policies. Introduced in the 1990s, EPR policies require producers to bear the financial responsibility for managing the disposal of their products, encouraging them to design items for more efficient and environmentally friendly disposal (Lifset et al., 2013). However, some scholars highlight how EPR policies face limitations in enforcement, dealing with 'ghost' products that are no longer identifiable and addressing gaps in globally dispersed secondary re-use markets (Arya & Bhutani, 2023).

EPR policies often overlook significant players in the tech supply chain, such as distributors and platforms, which play a crucial role in linking producers with consumers. This chapter contributes to the literature by focusing on an under-explored aspect of the tech supply chain: retailers. Specifically, it examines the role of 'retailer-platforms' as major consumers of digital hardware and as marketplaces that connect producers and consumers through global digital networks. Drawing from semi-structured interviews with retailers in the United States and Zimbabwe, this chapter compares retailers' perspectives in addressing e-waste across the Global North and Global South. The two regions were chosen because of the United States' centrality in the production and circulation of e-waste and Zimbabwe is understood as an exemplar of countries affected by the global flows of e-waste that are not so easy to track due to inter-regional flows between countries in the Global South.

Our analysis reveals how sociotechnical and economic path dependencies led to differently constructed retail markets across geographic contexts. The platformisation of retail in the United States was deeply influenced by the growth of industrial cloud companies combined with neoliberal economic policies, whereas post-colonial instability and economic transformation drove platformisation in Zimbabwe. The findings reveal how retailers in the Global North frequently shirk their responsibility in facilitating e-waste disposal by obscuring their recycling programmes, transferring the onus onto State governments with insufficient recycling infrastructure. In contrast, retailers in the Global South lack policies and recycling infrastructure, which places the burden of e-waste management on consumers and informal systems. Both regions rely on *platform opacity* – understood as the obfuscation of information related to e-waste management – which leads to low awareness of recycling programmes and the avoidance of responsibility in ensuring safe and equitable end-of-life disposal of electronics. Counter to Glissant's (1997) concept of opacity as a right for the marginalized, we critique the way tech retailers use opacity as a shield to evade accountability. We argue that the platformisation of retail has accelerated the e-waste crisis by facilitating increased consumption, exacerbated global inequities in the accumulation and circulation of e-waste and obscured the roles and responsibilities of retailers in addressing the disposal of consumer digital goods.

The chapter is composed of two sections: first, the chapter defines retailers within the platform economy and how they impact e-waste production and disposal. Second, we present findings from interviews with retailers and how they view their role in addressing the issue of e-waste, recycling programmes and disposal methods, and how retailers promote these services. To conclude, we propose a revised approach to *retailer-platform responsibility* to address the gaps in the allocation of responsibility across the tech supply chain.

The Platformisation of Retail in the United States and Zimbabwe

The platformisation of retail, defined as the transformation of markets into platforms that facilitate the exchange of goods and enable the accumulation of both data and economic assets through marketplace rents (Srnicek & De Sutter, 2017), has unfolded along radically different political economic trajectories in the United States and Zimbabwe. This has created e-waste regimes that appear opposite but are, in fact, interdependent. In the United States, financialisaton, neoliberal deregulation and corporate consolidation enabled the growth of the platform economy (Davis & Kim, 2015; Nieborg et al., 2022). In the early 2020s, online sales of consumer electronics in the United States accounted for 45.6% of all retail sales in this category, generating $218 billion in revenue and projected to grow to $285.5 billion by 2028 (ECDB.com, 2023). In contrast, post-colonial economic crisis in Zimbabwe led to underdeveloped payment structures, new local-grown retailers and the rise of mobile phone informal trading networks (Tsarwe & Mare, 2021). Yet both models ultimately feed the same global system of unequal e-waste distribution with overconsumption in the United States and informal reuse markets in Zimbabwe representing two sides of the same political economy of e-waste coin.

The Reciprocal Rise of E-Retail and the Cloud

The US model of retail platformisation reflects the broader political economy of financialised, neoliberal capitalism (Kotz, 2010). Amazon's rise as both retail platform and cloud computing provider was enabled by state policies favouring corporate consolidation – from tax subsidies to lax antitrust enforcement (Khan, 2017). However, the relationship between cloud computing and the platformisation of retail is more directly intertwined than a broad economic history might suggest. Instead, the fight to secure online retail market share precipitated in the development of cloud computing.

In the early 2000s, Amazon launched a platform service for retailers called merchant.com, which allowed retailers to build online shopping sites on top of Amazon's e-commerce platform (Miller, 2016). Amazon's attempt to capture the online retail space followed Microsoft's similarly named product, 'Merchant Server' (later Commerce Server), launched in the late 1990s, which gained considerable traction among United Kingdom retailers throughout the 2000s and early 2010s (Walker, 2011). Both companies sought to provide a computing platform solution for the retail sector, and as a result, the early commercial cloud was first conceived, providing the foundational infrastructure enabling the transformation of platform capitalism and retail in return.

To understand how the platformisation of retail enabled the increased consumption of electronics and, in turn, added to the production of e-waste, we need to unpack two simultaneous developments: (1) how the cloud impacted retail and (2) how retail impacted the cloud. First, an entirely new fleet of electronic products connected through the internet flooded the market under the auspice of the Internet of Things (IoT). Once devices were connected to the internet, inconspicuous household items such as refrigerators, doorbells and toothbrushes were reimagined as 'smart devices', serving primarily as portals to computing power in the commercial cloud. The IoTification of everyday objects contributes to e-waste in two ways: (1) through planned obsolescence and shoddy manufacturing of connected devices that are made to last a few short years (if that) and (2) by perpetuating to the endless data imaginary as described by Brown and Hogan (*this volume*), where toothbrushes and refrigerators now also produce endless streams of 'data waste' (Thylstrup, 2019). Additionally, the platformisation of retail has effectively expanded 'shelf space' infinitely, linking well-known retail brands and third-party sellers to new, geographically diverse customers. This shift has provided buyers with greater access to a wide array of products through online marketplaces. As long as retailers could coordinate logistics to move goods, they could sell even the most obscure items to distant customers, minimising the economic risks typically associated with brick-and-mortar retail, particularly if a product failed to sell.

Second, sociotechnical advancements in retail have driven up the demand for remote computing and reliance on cloud services. Retailers need robust computing infrastructure to stay competitive, as online electronics retail involves hyper-personalised advertising and the management of global logistics networks, all of which require stable and reliable computational power. This trend is likely to continue as more artificial intelligence (AI) tools are integrated into the retail sector. However, even if major retailers ceased using cloud computing, the electronics and IoT ecosystems are so intertwined with cloud computing that separating electronics sales from the cloud would require an entire overhaul of the electronics market.

Big cloud providers are known for their significant consumption of energy and water. They need large facilities, extensive land for new developments, and their operations also produce direct emissions from diesel generators (Bridges, 2024). What is sometimes overlooked in discussions of the environmental impacts of cloud infrastructure is how they also contribute millions of tons of e-waste. In 2020, one industry group (AKCP) estimated there were approximately 18 million servers in data centres around the world (up from 11 million in 2006), which need to be replaced on average every 2-3 years (Garcia, 2022). This means that each year, millions of servers enter the formal and informal trade of global e-waste. When these devices enter the informal market, they often end up in landfills located in the Global South, where e-waste workers are exposed to dangerous toxic metals and chemicals that have long-term environmental and health impacts (Maphosa & Mashau, 2023). Cloud infrastructures both enabled the platformisation of retail while increasing e-waste from direct consumption, and in turn, platformisation accelerated the e-waste crisis by facilitating increased consumption of electronics.

Electronics Retailers in Zimbabwe and Sub-Saharan Africa

Growing internet penetration, paired with the broad uptake of smartphones and expansion of mobile payment systems, has contributed to the platformisation of retail in Africa in recent years. Nigerian-based e-commerce and logistics company Jumia (colloquially called the Amazon of Africa) had over 32 million monthly visits in 2021, followed by South African e-retailer takealot.com and Amazon-owned online retailer Souq, with 10.5 million and 10 million visitors per month, respectively (Fleck, 2022). African online retailers also compete against Chinese-owned retail giants Shein and Temu and Killimall (Chinese start-up), which have been expanding operations in Northern Africa, and US-owned retailers Amazon and Walmart, which have been expanding operations in South Africa (Dludla, 2024). However, the growth of e-commerce is not consistent across Africa, with much more limited uptake of online shopping in East Africa and Sub-Saharan Africa partially due to lower internet connectivity, underdeveloped payment structures and low trust in online retail (Odusanya et al., 2022).

In Zimbabwe, the platformisation of retail has progressed slowly due to underdeveloped ICT infrastructure, poor economic conditions and a lack of regulations supporting e-commerce. Following the economic collapse in the late 1990s, major global ICT companies exited the market, allowing local firms to emerge and offer smaller-scale retail and ICT solutions. Axis Solutions and Dandemutande, two of the largest ICT retailers in Zimbabwe with a presence in

Sub-Saharan Africa, were founded during this period. Unlike in the United States, information on the market value, revenue and operations of Zimbabwean companies is challenging to find. Both companies provide ICT solutions, internet and cloud services, operating from their headquarters in Harare. Across the industry, online sales of consumer electronics in Zimbabwe reached around $43 million in 2023, accounting for 25.1% of e-commerce revenue (ECDB.com, 2023). While internet user penetration in Zimbabwe has reached around 42%, e-commerce was used by around 12.5% of the population in 2024 (eCommerce - Zimbabwe, 2024; US Department of Commerce, 2024). The Zimbabwean e-commerce market is expected to grow due to increased internet access and steady changes in consumer behaviour.

Logistically, retailers rely on ground transportation, rather than air freight, to move products within Zimbabwe and across borders. Porous national borders also mean that Zimbabwe receives a significant amount of used electronics via ground transportation coming from large port countries, which receive global shipments of used electronics. For example, Nigeria imports an estimated 60,000-71,000 tonnes of used electronics each year (Maes & Preston-Whyte, 2022). These containers are transported to landlocked countries like Zimbabwe, where some devices are reused for a short period before quickly becoming obsolete. The lack of policies and transparent e-waste management shifts the burden of disposal to informal recyclers. To encourage automation and participation in the knowledge economy, the Zimbabwean government has eliminated import duties on ICT devices, permitting institutions and individuals to import second-hand technology, which has further exacerbated the e-waste issue (Macherera et al., 2023).

Comparing the development of retail in the United States and Zimbabwe reveals the sociotechnical and political economic path dependencies that precipitated the modern and emerging e-commerce industry. Between the two nations, the United States maintains an asymmetrical trade relationship with Zimbabwe (and Africa more generally) in electronics and e-waste, under the guise of 'reuse' or 'donations' while avoiding accountability for hazardous waste disposal (Maphosa & Mashau, 2023). The platformisation of retail in the United States and growth of regional retail in Zimbabwe facilitated more ICT sales, while outpacing investments in e-waste recycling. The informal secondary use market plays an important role in Zimbabwe's electronic retail economy, and informal repair shops act as community hubs to fix old and refurbished electronics (Maphosa, 2021). While the trans-border movement of e-waste is illegal for most countries under the Basel Convention, the United States has not ratified the convention, which would require the passage of federal laws to ban the importation and exportation of e-waste (Singh, 2019). In 2022, the United States produced 7.2 MT of e-waste, more than the African continent (5.5 MT), while China led with 22 kg per capita, followed by the United States with 21 kg of e-waste per capita. Approximately 80% of the e-waste generated by the Global North is exported illegally to developing countries in the Global South (Yu et al., 2017). Despite global inequities, Africa is still a net exporter of e-waste (Lepawsky, 2018). This surprising statistic likely reflects systematic measurement flaws – misclassified 'used electronics' that goes unaccounted for such as those hidden in second-hand vehicle imports (Kuehr, 2018). Global disparities in electronics consumption mixed with insufficient reporting and enforcement perpetuate a form of 'waste colonialism' (Liboiron, 2021) wherein Africa serves as a market for obsolete electronics and a dumping ground for e-waste, while enabling accountability laundering through opaque supply chains and liability shifting for both consumers and tech giants located in the Global North.

Retailers' View of E-waste

To explore how retailers perceive their role and responsibility in addressing e-waste, we conducted interviews with electronics and ICT retailers in the United States and Zimbabwe. Because workers in managerial roles working for large retailers located in the United States did not respond to the researcher's request for interviews, covert 'mystery shopper' interviews were conducted with customer service agents and retailing staff over the phone ($N = 10$). Mystery shopper calling is a covert evaluation method where researchers pose as regular customers to assess the quality of customer service (Wilson, 1998). This method relies on realistic interactions to capture unfiltered employee communication, and it was used in this context to test customer service agents on their knowledge of company policies and practices in relation to handling e-waste. In Zimbabwe, semi-structured interviews were conducted with branch managers from two of the largest electronics and ICT retailers, Axis Solutions and Dandemutande ($N = 3$). These interviewees were recruited through LinkedIn, email and telephone, and the interviews were conducted via video conferencing. The semi-structured format allowed researchers to ask follow-up questions for more detailed responses while ensuring consistent themes across all interviews for comparability (e.g., e-waste knowledge and management practices, how customers recycle and attitudes about who should be responsible for e-waste processing). All interviews were then coded using abductive coding (Deterding & Waters, 2021) by first open-coding transcripts with issues that arose (e.g., disinterest in information about recycling), then refining thematic categories that were consistent across both sets of interviews (e.g. lack of awareness, opacity around programmes and deferring responsibility).

The disparity between the fragmented customer service interactions in the United States and the in-depth, semi-structured interviews with Zimbabwean branch managers reflects both political–economic and positional differences. Where frontline US agents provided scripted, programme-limited responses, Zimbabwean managers offered operational insights. This divergence underscores how corporate hierarchies mediate access to knowledge, as well as uneven fieldwork realities: one marked by algorithmic gatekeeping and the other by relational negotiations required for on-the-ground research. Following Bartlett and Velkova's (*this volume*) interdisciplinary and creative intervention to study data centres through analogue materials, we see this patchwork approach as a necessary tool to study opaque infrastructures, which evade easy interrogation. Rather than a flaw in our study, we see these tensions as productively revealing how institutional positionality filters understanding of e-waste and recycling programmes. Finally, as a collaboratively written chapter, authored from both the Global North and the Global South, we note that these productive disjunctures are also a product of institutional and epistemological inequities. We try to confront these disjunctures by amplifying marginalised scholars alongside dominant anti-colonial theorists, while ourselves practising a Glissantian relation by respecting difference with humility (see Glissant, 1997).

The findings reveal geographically specific responses to retailer e-waste policies and practices while also exposing the relationship *between* the two locations. Across the interviews with customer service agents and retail employees, there were several notable themes: 1) information about existing recycling programmes was difficult to locate either due to a lack of awareness by staff or these programmes were known but not actively promoted; 2) information about where electronics go to be recycled and which partners the corporations work with was unknown or obscured; and 3) many retailers deferred responsibility for handling e-waste by directing customers to third-party locations or state-run programmes, some of whom caught illegally dumping, while others ultimately placed the responsibility on the customer to dispose of their electronics.

Awareness and Impossibility of Studying E-waste Management in Retail

In Zimbabwe and the United States, poor information management around recycling programmes and limited knowledge on the environmental impacts of e-waste created barriers for studying platform e-waste management. In the United States, it took two or three attempts to reach a customer service agent or retail employee to discuss the electronics recycling programmes. It was clear that the automated customer service systems were designed to respond to customers with an existing order, and there were several barriers facing someone seeking more information about company programmes, like recycling. For example, when the researcher called Amazon's customer service line, after a long back-and-forth with the automated system, they were transferred to a human agent. When they asked for information about the recycling programme, the customer service agent responded with surprise, 'Recycling! Let me do some research on that quickly'. Then the line went dead.

On the second attempt (Interview, 22 July 2024), the researcher again navigated through the automated messages to another human agent and asked a series of questions. Then, the following exchange took place:

- Agent: Can you provide your email associated with your Amazon account? I will email you some information on the recycling programme [...].
- Researcher: I don't have an Amazon account.
- Agent: Do you have a phone associated with your Amazon account?
- Researcher: No, ma'am [...].

- Agent: Ma'am, I will just give you a website, and you can search for it on your end.
- Researcher: So, you can't send me information on the recycling programme if I don't have an Amazon account?

– Line went dead –

While Amazon and Apple have expansive recycling programmes, they are not widely known about or actively promoted. Searching for 'Amazon electronics recycling' reveals that recycling with Amazon is completely free. The company provides a trade-in service for devices that still have market value and offers options to mail in unwanted electronics with a free shipping label. Additionally, customers can drop off electronics at local Staples stores or third-party organisations. Similarly, Apple recycles any of its products. However, when asked about the volume of recycling at an Apple store, an employee noted, 'It's not excessive. We don't get products every day because not a lot of people know about the recycling programme' (Interview, 19 July 2024).

Interviews with retailers in Zimbabwe, like those in the United States, revealed a lack of transparency regarding e-waste management. However, Zimbabwean retailers focused more on the need for public awareness and education about the dangers of e-waste, the underdeveloped recycling infrastructure and the responsibility of retailers in facilitating e-waste collection and disposal. Since electronics recycling is largely informal in Zimbabwe, retailers lack established partnerships with third-party recyclers compared to the United States, leading to no formal e-waste recycling programmes. One worker noted that general waste management is a broader issue in Zimbabwe: 'If you look around, you find that we do not even have enough trash cans just to collect municipal waste. You know, just the ordinary litter that we come across every day' (Interview, 8 August 2024).

Zimbabwean retail workers also expressed concern about customers' lack of awareness regarding the potential harms of e-waste. One interviewee noted, 'We need to raise awareness about the hazards associated with improper waste disposal so that consumers follow the correct procedures for e-waste disposal. For instance, some might consider burning e-waste, such as television sets' (Interview, 8 August 2024). The worker stressed that retailers should educate customers on proper e-waste handling by providing warnings and guidelines on products. They also highlighted the need for government bodies, including consumer organisations and environmental management agencies, to participate in public education and the development of e-waste policies.

Opacity Around Flows of E-waste

While retail-led e-waste recycling programmes are common among some of the top e-retailers in the United States, there is little to no information about where e-waste flows after it is returned to the retailer, which may increase the risk of e-waste entering the informal market. E-retailers with recycling programmes sometimes named third-party partners on their websites though information on partner practices was difficult to locate. Several platforms did not identify which third-party they work with, and customer service agents did not have this information (see Table 7.1). For example, because both HP and Dell work with third-party providers – Goodwill and Best Buy – the customer service agents eventually directed the researcher to these organisations for more information. The lack of information in the United States increases the risk of mismanagement of e-waste recycling and makes it difficult to hold platforms accountable for third-party practices.

The e-waste and electronic recycling industry in Zimbabwe is still underdeveloped, leaving both customers and retail companies unsure of how to manage electronics at the end of their useful life. Similar to the situation in the United States, Zimbabwean retailers lack information about recycling programmes and proper e-waste disposal methods. One retail worker from Axis Solutions shared, 'We recently replaced our printers with new models, and the old ones are just stored in a room. I wouldn't know the best way to dispose of them'. A Zimbabwean retail worker highlighted the lack of proper e-waste management, stating that people dispose of e-waste in the same way as domestic waste, simply placing it in trash cans (Interview, 8 August 2024). Another retail worker explained, 'There is no one I know who is involved with handling this waste' (Interview, 23 July 2024).

Deferred Responsibility

There was a common theme across interviews of retailers deferring responsibility for e-waste management to another third-party or to some indeterminate time in the future. Three of the top 10 online electronics retailers in the United States – Walmart, Sam's Club and Costco – do not offer electronic recycling programmes. When one of the authors contacted Walmart, a customer service agent referred them to a third-party website, mrmrecycling.com, for local recycling options. Similarly, Costco directed the researcher to state government websites for information on e-waste handling and state-run programmes (see Table 7.1). For retailers that do offer recycling programmes, employees often lacked knowledge about their partnerships or the final destination of the recycled items. For example, an Apple store employee stated, 'I don't know exactly where they end up, but they are taken into parts. But products aren't resold or anything' (Interview, 19 July 2024). A Target worker mentioned that they only have a small bucket for e-waste, which is sent to an outside vendor, but they had no information about that vendor.

The research highlighted inconsistencies in responses from different retailers. For instance, a Walmart representative directed the researcher to a third-party website for recycling locations, while Costco referred them to state government resources. Target accepted small electronics but revealed a lack of transparency regarding their recycling process, with items sent to an unknown outside vendor.

Table 7.1. Electronic Recycling Programmes for Top Online Electronics Retailers (2024).

Retailer	Recycle Programme?	Cost?	Which Products?	Managed by
Amazon.com	Yes	No	All electronics	Re-Tech part of the Li Tong Group based in Hong Kong
Apple.com	Yes	No	All Apple products	Li Tong Group, Hong Kong; Brightstar Corp., FL; TES-AMM, Singapore; and Foxconn Technology Group, Taiwan (Culpan, 2016)
Walmart.com	No	NA	NA	NA
Bestbuy.com	Yes, limit 3/HH/day	Yes, for larger items	Electronics with no batteries	Dynamic Lifecycle Innovations, WI; Electronic Recyclers International (ERI), CA; Regency Technologies, OH; Universal Recycling Technologies, LLC (URT), WI
Dell.com	Yes	No	All Dell products	Goodwill
Target.com	Yes	No	Small items	Unknown
Samsclub.com	No	NA	NA	NA
Costco.com	No	NA	NA	NA
Ring.com	Yes	No	All	Through Amazon
Hp.com	Yes	Drop-off	HP products	In the U.S. they partner with BestBuy for drop off. HP work with 44 different partners located all over the world (HP Reuse and Recycling Vendors, 2024).

This lack of clarity is concerning, especially given Target's history of illegal hazardous waste disposal, including a $22.5 million settlement for dumping in California in 2011 and a $7.4 million fine in 2018 for e-waste violations (Haskins, 2018; Li, 2011). Similarly, in 2021, Walmart faced legal action for illegally dumping hazardous materials, including lithium batteries, e-waste and confidential customer information, in California landfills (Martin, 2021). The situation is compounded by the fact that only 23 states and the District of Columbia have landfill bans specifically prohibiting e-waste disposal, creating a gap in accountability for retailers.

The study also contrasts this with insights from retailers in Zimbabwe, where awareness and education about e-waste are severely lacking. The informal sector largely handles e-waste, and retailers do not have established recycling programmes like their US counterparts. Retail workers indicated a significant gap in waste management infrastructure, with many simply disposing of e-waste alongside regular trash. Workers expressed a need for increased public awareness about the dangers of improper e-waste disposal. They suggested that retailers should take on an educational role, providing warnings and guidelines for consumers on how to handle e-waste. There was also recognition of the necessity for government involvement in developing policies and educational initiatives regarding e-waste management. Despite the challenges, retail workers acknowledged that broader waste management issues in Zimbabwe hinder effective e-waste handling. Many pointed out a lack of basic waste collection facilities, complicating efforts to implement any formal e-waste recycling processes. Additionally, there was frustration regarding ICT companies that do not accept back obsolete electronics, effectively shifting the disposal burden to consumers, which raises safety concerns, particularly with malfunctioning batteries.

The Zimbabwean retailers had differing views on who should be ultimately responsible for discarding e-waste. One retail worker suggested that producers should take responsibility for discarding their products because they are the ones making and selling devices to retailers, they explained 'ICT companies [producers] are the ones selling electronic gadgets. But they don't take back obsolete electronics. The ICT companies keep pushing the burden to the consumer' (Interview, 23 July 2024). The Zimbabwean retail worker outlined a programme akin to EPR policies, which would hold producers accountable for the disposal of electronics. However, another retail worker argued that retailers should also take responsibility for assisting customers in discarding unwanted electronics. They stated, 'Since we are the ones distributing the equipment to the public, we should be at the forefront of managing e-waste and implementing policies' (Interview, 8 August 2024).

How Platform Opacity Exacerbates Global Disparities in Retail Responsibility

Understanding the ways in which platform opacity impacts global e-waste flows and disparities is more than an issue of making information transparent.

A subverted reading of Glissant's (1997) concept of the 'right to opacity' is useful here. Where Glissant's opacity empowers the marginalised by resisting colonial demands for explanation, platform-retailers deploy opacity asymmetrically to obscure responsibility in the global e-waste issue, while evading accountability through opaque global supply chains. Following this line of critique, Renteria (2024) argues we instead need to interpret opacity as a problem of *information management* that goes beyond the mere asymmetries of information and instead moves us towards transparency and accountability. The interviews with electronics retailers in the United States and Zimbabwe reveal significant disparities in the development and accessibility of electronics recycling infrastructures. In the United States, many major online platform-retailers have created recycling programmes through partnerships with third-party providers. However, information about these programmes is often hard to locate and may even be deliberately obscured. There is also a general lack of transparency concerning how items are recycled, which companies are involved, and the fate of items once returned to retailers.

Despite benefiting from electronics sales, platform-retailers seem to struggle with their responsibility in addressing e-waste disposal adequately. For instance, a journalist noted in 2022 that they recently discovered Amazon's recycling programme, which had existed for some time but was unknown to many (Roth, 2022). Interviews further revealed that even when recycling programmes are available, many customers do not utilise them due to a lack of awareness. Additionally, when customers attempt to inquire about recycling programmes by speaking with retail staff, they are likely to encounter several barriers, making it challenging to answer basic questions on what to do with their electronics they wish to discard. This lack of understanding opens the door for retailers to engage in questionable practices without significant public oversight. Without national regulations on e-waste dumping, it is effectively legal in many states for retailers to dispose of their e-waste into landfill. However, even when there are legal protections in place, large retailers have been caught flouting regulations, as seen with the example of Target and Walmart illegally dumping e-waste in California. The compounding layers of platform opacity in the United States increase the risk that improperly handled e-waste will end up in the informal waste streams and eventually make its way from the Global North to the Global South, and countries like Zimbabwe where there are few barriers to receiving unusable electronics and few protections for workers and communities dealing with the toxic exposure from e-waste.

Conclusion: Towards Platform-Retail Responsibility

In this chapter, we examined how cloud computing has enabled the platformisation of retail in the United States, while Zimbabwe's economic collapse in the late 1990s fostered the growth of local ICT businesses that are still in the early stages of adopting e-commerce. The platformisation of retail has intensified the e-waste crisis by promoting increased consumption and worsening global inequities in the handling of discarded electronics. A lack of accountability in both regions contributes to rising e-waste volumes, complicating sustainable recycling and disposal efforts. Both the United States and Zimbabwe experience a shortage of information and transparency regarding e-waste management, increasing the risk of mismanagement and creating barriers to accountability. In the United States, navigating multiple automated systems hinders access to recycling information, while in Zimbabwe, the absence of policies and infrastructure shifts the responsibility for e-waste management onto unknown producers in the Global North or to an uncertain future.

To address the limitations of existing EPR policies, we propose a new practical approach called platform-retailer responsibility (PRR). This approach acknowledges the dual role of retailers as both consumers and distributors of electronics. Under PRR, retailers could be held economically responsible for facilitating the return and recycling of the products they sell. This could involve contributing a portion of their sales revenue to local or national recycling initiatives and incentivising awareness of their recycling programmes through specific recycling goals. Retailers lacking free recycling options could face fines, which would need to be substantial enough to motivate the development of reverse logistics networks.

We suggest that these programmes could complement EPR policies, allowing producers to pay retailers to enhance their reverse logistics and connect customers with recyclers. This platform-responsibility practical approach aims to better address global disparities in electronic consumption and e-waste production, assigning partial responsibility to the platforms involved in selling electronics. Future research is essential to further analyse the contributions of platform-retailers to the e-waste problem, both as consumers and sellers of electronics. Additionally, more investigation is needed into the limitations of existing EPR policies, particularly regarding the transfer of used electronics to secondary markets that may not adhere to regional or national regulations. Solutions must involve collaboration among policymakers, industry leaders and communities affected by e-waste management. Addressing the global e-waste issue requires not only international cooperation, such as treaties like the Basel Convention, but also a focus on disparities in consumption patterns and financial flows. Ultimately, we emphasise the need for accountability among different actors within the same industry across various countries in contributing to the global waste problem.

References

Arya, A., & Bhutani, A. (2023). *Study on items shipped for reuse and extended producer responsibility fees: A case for extending EU EPR fees to cover end-of-life activities of products shipped outside the EU* (pp. 1–67). European Environmental Bureau. https://eeb.org/wp-content/uploads/2023/05/items-shipped-for-reuse-and-EPR-fees_EEB-2023-.pdf

Baldé, C. P., D'Angelo, E., Luda, V., Deubzer, O., & Kuehr, R. (2022). *Global transboundary e-waste flows monitor 2022*. United Nations Institute for Training and Research (UNITAR).

Baldé, C. P., Kuehr, R., Yamamoto, T., McDonald, R., D'Angelo, E., Althaf, S., Bel, G., Deubzer, O., Fernandez-Cubillo, E., Forti, V., Gray, V., Herat, S., Honda, S., Iattoni, G., Khetriwal, D. S., Luda di Cortemiglia, V., Lobuntsova, Y., Nnorom, I., Pralat, N., & Wagner, M. (2024). *The global e-waste monitor 2024*. United Nations Institute for Training and Research (UNITAR). https://ewastemonitor.info/the-global-e-waste-monitor-2024/

Bartlett, F. M., & Velkova, J. (this volume). This photograph has never been in a data centre: Critical experimental interventions in the pursuit of sustainable research practice. In O. Dovbysh, S.-M. Laaksonen, & M. Pantti (Eds.), *Digital platforms and environmental responsibility*. Emerald Publishing Limited.

Bridges, L. E. (2024). Competing digital capacities: Between state-led digital governance and local data center tradeoffs. *Information, Communication & Society*, *0*(0), 1–18. https://doi.org/10.1080/1369118X.2024.2331765

Brown, T. J., & Hogan, M. (this volume). Synthetic DNA, sustainable computing and the data hoarding crisis. In O. Dovbysh, S.-M. Laaksonen, & M. Pantti (Eds.), *Digital platforms and environmental responsibility*. Emerald Publishing Limited.

Culpan, T. (2016). *Where your iPhone goes to die (and be reborn)*. https://www.bloomberg.com/news/features/2016-02-16/where-your-iphone-goes-to-die

Davis, G. F., & Kim, S. (2015). Financialization of the economy. *Annual Review of Sociology*, *41*, 203–221. https://doi.org/10.1146/annurev-soc-073014-112402

Deterding, N. M., & Waters, M. C. (2021). Flexible coding of in-depth interviews: A twenty-first-century approach. *Sociological Methods & Research*, *50*(2), 708–739. https://doi.org/10.1177/0049124118799377

Dludla, N. (2024). *Amazon is in South Africa for the long run, says marketplace head*. Reuters. https://www.reuters.com/business/retail-consumer/amazon-is-south-africa-long-run-says-marketplace-head-2024-09-12/

ECDB.com. (2023). *Consumer electronics ecommerce market in U.S.* https://ecommercedb.com/markets/us/consumer-electronics

eCommerce—Zimbabwe. (2024). *Statista*. https://www.statista.com/outlook/emo/ecommerce/zimbabwe

Fleck, A. (2022). *Infographic: Africa's e-commerce giants*. Statista Daily Data. https://www.statista.com/chart/27527/africas-popular-e-commerce-platforms

Gabrys, J. (2013). *Digital rubbish: A natural history of electronics* (First paperback ed). University of Michigan Press.

Garcia, C. (2022). *The amount of data center energy use—AKCP monitoring. AKCP remote sensor monitoring*. Data Center Monitoring. https://www.akcp.com/blog/the-real-amount-of-energy-a-data-center-use/

Glissant, É. (1997). *Poetics of relation* (B. Wing, Trans.). University of Michigan Press.

Haskins, C. (2018). *Target fined $7.4 million for illegally throwing e-waste in the trash*. Vice. https://www.vice.com/en/article/target-fined-74-million-for-illegally-throwing-e-waste-in-the-trash/

HP reuse and recycling vendors. (2024). https://h20195.www2.hp.com/V2/GetDocument.aspx?docname=c05403198

Khan, L. M. (2017). Amazon's antitrust paradox. *The Yale Law Journal, 126*, 710–805. https://scholarship.law.columbia.edu/faculty_scholarship/2808

Koebler, J. (2016). *A shocking amount of e-waste recycling is a complete sham*. Vice. https://www.vice.com/en/article/aeky44/much-of-americas-e-waste-recycling-is-a-sham

Kotz, D. M. (2010). Financialization and neoliberalism. In y G. Teeple, & S. McBride (Eds.), *Relations of global power* (pp. 1–18). University of Toronto Press.

Kuehr, R. (2018). *Person-in-the port (PiP) study*. SCYCLE. https://www.scycle.info/person-in-the-port-pip-study/

Lepawsky, J. (2015). The changing geography of global trade in electronic discards: Time to rethink the e-waste problem. *The Geographical Journal, 181*(2), 147–159. https://doi.org/10.1111/geoj.12077

Lepawsky, J. (2018). *Reassembling rubbish: Worlding electronic waste*. The MIT Press.

Li, S. (2011). *Target to pay $22.5 million to settle hazardous-waste dumping case*. Los Angeles Times. https://www.latimes.com/local/la-xpm-2011-feb-04-la-me-target-settlement-20110205-story.html

Liboiron, M. (2021). *Pollution is colonialism*. Duke University Press.

Lifset, R., Atasu, A., & Tojo, N. (2013). Extended producer responsibility. *Journal of Industrial Ecology, 17*(2), 162–166. https://doi.org/10.1111/jiec.12022

Macherera, M., Maphosa, V., & Zezai, D. (2023). Perceptions and practices on e-waste management in selected institutions of Zimbabwe. *Water and Environmental Sustainability, 3*(2), 22–31.

Maes, T., & Preston-Whyte, F. (2022). E-waste it wisely: Lessons from Africa. *SN Applied Sciences, 4*(3), 72. https://doi.org/10.1007/s42452-022-04962-9

Maphosa, V. (2021). E-waste management and practices at Zimbabwe's higher education institutions. *Journal of Higher Education Theory and Practice, 21*(1), 155–166.

Maphosa, V., & Mashau, P. (2023). The conundrum: Transforming African e-waste landfills to urban mines. In H. M. Saleh, A. I. Hassan, & R. F. Aglan (Eds.), *Advances and challenges in hazardous waste management*. IntechOpen. https://doi.org/10.5772/intechopen.1002419

Martin, H. (2021). *Walmart sued for alleged dumping of hazardous waste in California landfills*. Los Angeles Times. https://www.latimes.com/business/story/2021-12-20/walmart-california-lawsuit-alleges-hazardous-waste-dumping-landfills

Miller, R. (2016). *How AWS came to be*. TechCrunch. http://social.techcrunch.com/2016/07/02/andy-jassys-brief-history-of-the-genesis-of-aws

Nieborg, D., Poell, T., & van Dijck, J. (2022). Platforms and platformization. In T. Flew, J. Thomas, & J. Holt (Eds.), *The SAGE handbook of the digital media economy*. SAGE Publications. https://www.torrossa.com/en/resources/an/5409560

Odusanya, K., Aluko, O., & Lal, B. (2022). Building consumers' trust in electronic retail platforms in the sub-saharan context: An exploratory study on drivers and impact on continuance intention. *Information Systems Frontiers, 24*(2), 377–391. https://doi.org/10.1007/s10796-020-10043-2

Plantin, J.-C., Lagoze, C., Edwards, P. N., & Sandvig, C. (2018). Infrastructure studies meet platform studies in the age of Google and Facebook. *New Media & Society, 20*(1), 293–310. https://doi.org/10.1177/1461444816661553

Renteria, C. (2024). Opacity as an information problem: A framework for refreshing the study of opacity. In O. Meza, & E. Pérez-Chiqués (Eds.), *The structure of systemic corruption* (pp. 69–92). Springer. https://doi.org/10.1007/978-3-031-64117-6_4

Roth, E. (2022). *Today I learned Amazon will recycle small electronics for free*. Verge. https://www.theverge.com/2022/7/16/23205725/today-i-learned-til-amazon-recycle-small-electronics-free

Singh, A. (2019). Out of sight, out of mind: How the United States discards e-waste. *Berkeley Political Review*. https://bpr.studentorg.berkeley.edu/2019/12/05/out-of-sight-out-of-mind-how-the-united-states-discards-e-waste/

Srnicek, N., & De Sutter, L. (2017). *Platform capitalism*. Polity.

Thylstrup, N. B. (2019). Data out of place: Toxic traces and the politics of recycling. *Big Data & Society*, *6*(2), 2053951719875479. https://doi.org/10.1177/2053951719875479

Tsarwe, S., & Mare, A. (2021). Mobile phones, informal markets and young urban entrepreneurs in Zimbabwe: An exploratory study. *Area Development and Policy*, *6*(3), 347–362. https://doi.org/10.1080/23792949.2020.1790021

U.S. Department of Commerce. (2024). *Zimbabwe—eCommerce*. International Trade Administration. https://www.trade.gov/country-commercial-guides/zimbabwe-ecommerce

Walker, B. (2011). *Microsoft folds its hand and abandons commerce server: What it means*. Brian Walker's Blog. https://web.archive.org/web/20150103224239/http://blogs.forrester.com/brian_walker/11-11-11-microsoft_folds_its_hand_and_abandons_commerce_server_what_it_means

Wilson, A. (1998). The role of mystery shopping in the measurement of service performance. *Managing Service Quality*, *8*(6), 414–420. https://doi.org/10.1108/09604529810235123

Yu, E. A., Akormedi, M., Asampong, E., Meyer, C. G., & Fobil, J. N. (2017). Informal processing of electronic waste at Agbogbloshie, Ghana: Workers' knowledge about associated health hazards and alternative livelihoods. *Global Health Promotion*, *24*(4), 90–98.

Section 3

Discourses of Responsibility

Chapter 8

Counter Accounts and Digital Advocacy: How Activists Enforce Platform Companies' Environmental Accountability

Markus Ojala and Mervi Pantti

University of Helsinki, Finland

Abstract

Platform companies have become targets for climate activism due to their environmental impact, high visibility and economic clout. This chapter theorises climate activists as key drivers of corporate environmental accountability and identifies counter-accounting and digital advocacy as prominent practices in these efforts. The authors posit that platform companies present a complex and challenging target for these practices, as climate activists rely heavily on the very platforms they criticise to gain visibility for their accountability claims. In this empirical study, the authors examine the use of counter accounts and digital advocacy by environmental non-governmental organisations (ENGOs) to hold platform companies accountable for climate damage. By examining ENGO campaigns since 2009, the authors identify four key issues that climate activists focus on in their accountability practices targeting platform companies: (i) energy consumption and transition to renewable energy, (ii) climate policy advocacy, (iii) cooperation with fossil fuel companies and (iv) climate disinformation. The authors conclude by discussing how these findings highlight the complexities involved in how climate activism approaches platforms.

Keywords: Climate activism; environmental non-governmental organisations; platform companies; accountability; counter account; digital advocacy

Platforms and the Planet, 135–151

doi:10.1108/978-1-83662-172-020261009

Introduction

On a chilly February morning in 2019, Jesse Boateng, Greenpeace USA's Engagement Director, stands in front of the Crystal City station in Arlington, Virginia, the site where Amazon had recently announced it would build its new headquarters (Greenpeace USA, 2019). Wearing a dark green Greenpeace windbreaker and holding a microphone in his hand, Boateng is being filmed stopping commuters and asking them to pose questions to a human-sized Alexa standing beside him. As hesitant passers-by present questions about Amazon's energy consumption, a female voice from inside the mock Alexa makes witty remarks about the company's failure to meet their renewable energy commitments. Alexa confesses to a passer-by that she feels 'so dirty' for being powered by coal and gas.

The video featuring the mock Alexa video was shared on Greenpeace USA's YouTube and other social media channels. The captions explain the motivation behind the stunt: 'In 2014, Amazon committed to using 100% renewable energy to power its cloud. Since 2017, Amazon has built 23 new data centres in Virginia, but ZERO new renewable energy projects'. The stunt aimed to increase pressure on the company to uphold its commitments and generate awareness for Greenpeace's Click Clean campaign that monitors tech companies' energy use and investments in renewable energy. Greenpeace criticised the state's main electricity provider, Dominion, for a slow transition to renewable energy sources. However, the campaign did not target power companies directly, instead it applied pressure through platform companies due to their high visibility and perceived impact on utility companies' investment decisions.

Greenpeace's activism can be located within a broader trend of subjecting digital platform companies to public pressure in an effort to enforce their environmental accountability. The Alexa video represents an act of counter-accounting that resists Amazon's 'official' environmental discourse. In today's information landscape, the success of non-profit campaigns often hinges on the production of such counter accounts – that is, critical assessments of corporate conduct (She, 2022) – as well as digital advocacy, which is the creation social media content that provokes strong user reactions and attracts the attention of mainstream journalism (e.g., Blum & Uldam, 2024; Karpf, 2020). However, platform companies present a challenging target for these practices for two main reasons. First, due to their central role in almost all industrial sectors, platform companies are involved in climate change in various ways, making it difficult to prioritise campaigns for activism. Second, platform companies own and control the main online spaces that climate activists depend on to amplify their calls for accountability (Kavada et al., 2023).

This chapter explores whether and how the complexities mentioned above are evident in the selection and development of issues raised by climate advocacy groups in their efforts to hold platform companies accountable for their lack of action on climate change. We examine the utilisation of counter accounts and digital advocacy by international environmental non-governmental organisations (ENGOs) in their campaigns that target platform companies, identifying recurring themes and arguments. The main emphasis of our analysis is on the accountability claims made by ENGOs, specifically the issues for which they hold platform companies accountable.

The next section presents an overview of the role of ENGOs in holding corporations accountable for climate change impacts and introduces the concepts of counter-accounting and digital advocacy. This is followed by an analysis of ENGOs' efforts to enforce the accountability of platform companies in respect of (i) transition to renewable energy, (ii) climate policy advocacy, (iii) business cooperation with fossil fuel companies and (iv) climate disinformation and the online harassment of environmental activists. In conclusion, we discuss how these issues demonstrate the complexities involved in how climate activism aims to enforce platform companies' environmental accountability.

Climate Activists as Enforcers of Platform Companies' Environmental Accountability

Activists and NGOs are recognised as key agents of accountability that drive corporations towards greater societal responsibility (Coombs & Holladay, 2015; Maktoufi et al., 2020; Newell, 2008). High-visibility campaigns can impact corporate reputations and pressure companies to modify their behaviour (King, 2008; King & McDonnell, 2015). As 'accountability agents' (Moore, 2014) – that is, enforcers of companies' environmental accountability – ENGOs monitor corporate performance in meeting climate targets. The success of environmental campaigns often depends on their ability to garner support from movement partners, opinion leaders, the news media and various segments of the public (Cox, 2013; Damgaard & Lewis, 2014).

Since the 1980s, environmental organisations and activists have campaigned against fossil fuel companies and other industries that are historically responsible for high greenhouse emissions. Platform companies have become targets of environmental activism in recent years due to their political influence, economic clout, public visibility and growing climate impact. Nonetheless, only a few studies have addressed activist campaigns directed at platform companies (e.g., Katz-Kimchi & Manosevitch, 2015; Vaughan et al., 2023). In many ways, platform companies constitute a particularly challenging target for climate activism. First, due to their central role in almost all industrial sectors, platform companies are involved in climate change in various ways. Their impact on climate extends well beyond the emissions caused by their own operations and is related to the types of extractive industries and consumption they promote (see Laaksonen & Premsankar, 2026, this volume; Vigren et al., 2026, this volume). Given the complexity of the issue and the limited resources available to climate activists, ENGOs must make difficult choices regarding which issues to prioritise in their efforts to enforce corporate environmental accountability. Second, as providers of social media services, platform companies play a central role in directing public attention and facilitating digital advocacy (Blum & Uldam, 2024; Guo & Saxton, 2018). At the same time, social media has become a channel for climate misinformation and disinformation (Cook, 2019; Sethi, 2024), as well as a platform for the harassment of progressive activists.

Counter Accounts and Digital Advocacy

The production of counter accounts – that is, alternative representations of organisational practices and impacts – is a key activist tactic to make corporate activities more visible (She, 2022; Vinnari & Laine, 2017). These accounts can take various forms, including reports, audits, videos, blog posts, maps and social media messages (Saxton & Waters, 2014; She, 2022). Counter accounts are driven by the perceived failure of formal governance and accountability practices to guarantee protection of the environment and human rights (Moerman & van der Laan, 2015; Thomson et al., 2015). Often produced by NGOs, counter accounts aim to report on harmful corporate social or environmental practices (Denedo et al., 2017; Dey, 2003). They provide a counter-narrative to powerful representations such as mandatory corporate sustainability reporting, which highlight positive achievements for the purpose of legitimacy building. By spreading information on harmful corporate activities, counter accounts aim to mobilise public action and influence the decision-making of companies.

While they may fail to influence corporate conduct, counter accounts are acknowledged to possess the potential to raise awareness, diversify the public discussion on corporations' conduct and shed light on the gaps in environmental governance that enable corporations to maintain destructive practices (Denedo et al., 2017; Rodrigue, 2014). Moreover, the counter-accounting practices of ENGOs may give a voice to local communities and oppressed groups that suffer from the consequences of climate change, pollution, and biodiversity loss but are often sidelined in corporate reporting (Denedo et al., 2017; Vinnari & Laine, 2017). While these indirect avenues of influence are difficult to trace, there are indications that activist campaigns can influence corporate sales, stock prices and reputation (Dupire et al., 2022). Consequently, campaigns may lead to improvements in companies' conduct even though they may fall short of activist expectations (Davis et al., 2022; McDonnell et al., 2015).

The success of environmental campaigns typically hinges on their ability to generate support among movement partners, opinion leaders, the news media and various segments of the public (Cox, 2013). Due to the prevalence of digital networks and social media, activist campaigns increasingly involve digital advocacy (Johansson & Scaramuzzino, 2019; Karpf, 2020), that is, efforts to catch the attention of online publics and to mobilise users to join events, appeals or boycotts (Katz-Kimchi & Manosevitch, 2015). Social media platforms have become key channels for digital advocacy and the distribution of counter accounts (Saxton & Waters, 2014). The importance of reaching online publics means that environmental activism is increasingly reliant on social media to mobilise public support and to facilitate debates regarding corporate accountability.

Environmental Campaigns Targeting Platform Companies

The field of environmental activism is extensive and diverse, ranging from large international ENGOs to local networks of activists. We concentrated on prominent US-based and international ENGOs: Greenpeace, Friends of the Earth, Sierra Club, 350.org, Climate Action Against Disinformation, Global Action Plan, Natural Resources Defence Council, Sunrise Movement, Avaaz, ClimateVoice and Global Witness. Organisations' webpages were searched for the presence of digital advocacy and counter accounts targeting platform companies by combining the websites' own search functions and Google Search. Organisations that did not target platform companies in their activities were removed. However, most of these ENGOs had engaged in climate-related activism directed at platform companies. The collected data included all located reports and scorecards, as well as campaign websites and blog posts that addressed platform companies' environmental performance (Table 8.1).

In the analysis, we observed how ENGOs' use of counter accounts and digital advocacy directed attention to specific issues or problems in the conduct of platform companies. Individual issues were then aggregated into broader categories. Through our analysis, we identified four distinct focus areas in ENGOs' campaigns: (i) driving the transition to renewable energy, (ii) pushing for stronger climate policy advocacy, (iii) denouncing business cooperation with fossil fuel companies and (iv) demanding action to mitigate climate change disinformation and the online harassment of climate experts and activists. The following section presents each category, together with examples of digital advocacy and counter-accounting practices in each case.

Driving the Transition to Renewable Energy

Greenpeace has been instrumental in communicating climate change to the public since the 1990s (Doyle, 2007). While initially targeting the fossil fuel industry, Greenpeace has since launched several campaigns to push platform companies to reduce their emissions and to drive the shift to renewable energy. Its activities in this area intensified in 2009 with the launch of the Cool IT campaign that called for information and communication technology (ICT) companies to be 'strong advocates for policies that combat climate change and increase the use of renewable energy' (Greenpeace, 2010, p. 10). As part of the campaign, Greenpeace began a sustained counter-accounting effort by publishing an annual Cool IT Leaderboard (2009–2013) that ranked companies based on three criteria: how they offered technological solutions to end climate change, how they reduced their emissions, and how they engaged in political advocacy to advance the transition to renewable energy. In 2011, Greenpeace shifted its focus specifically to platform companies and cloud computing. The 'How Dirty Is Your Data?' report introduced a new company scorecard that evaluated the performance of the top 10 platform and cloud companies, based on the carbon footprint of their data centres (Greenpeace, 2011). The group continued its counter-accounting efforts in subsequent reports, citing the rapid growth in

Table 8.1. Breakdown of the Data.

ENGO	Reports and Scorecards	Campaign Blogs & Websites
350.org	–	2
Avaaz	1	4
Climate Action Against Disinformation	2	3
ClimateVoice	–	2
Friends of the Earth	1	5
Global Action Plan	–	2
Greenpeace	19	4
Global Witness	3	2
Natural Resources Defence Council	–	4
Sierra Club	–	3
Total	**25**	**31**

internet data traffic and the associated demand for energy, while advocating the acceleration of investments in renewable energy.

As platform companies' stated ambitions on renewable energy steadily grew throughout the 2010s, Greenpeace's counter-accounting and digital advocacy focused on the disconnect between promises and actions (Greenpeace, 2017). Accordingly, Greenpeace advocated that platform companies use their leverage to pressure electricity providers to increase the supply of renewable energy (Greenpeace, 2019). In the 2020s, the group's digital advocacy extended its geographical scope as Greenpeace East Asia began to issue yearly reports tracking the progress of leading China-based cloud service companies towards renewable energy (Greenpeace, 2020a). By 2024, these counter accounts targeted 25 leading cloud providers and data centre operators in China, noting slow but steady progress both in the companies' climate commitments and in their actual procurement of renewable energy (Greenpeace, 2024).

Pushing for Stronger Climate Policy Advocacy

Another front in climate activism has focused on the political advocacy of platform companies. Recognising the political power of platform companies, activists have demanded that they use their leverage to drive governmental policy change towards increasingly ambitious emission reduction targets and investments in renewable energy. As noted in the previous section, Greenpeace's counter accounts included evaluations of platform companies' climate policy advocacy. Another key actor in this area has been ClimateVoice, a group specifically dedicated to 'directing the massive influence of companies to help address the climate crisis' (ClimateVoice, 2024a).

In contrast to Greenpeace, ClimateVoice's digital advocacy is directed primarily at the employees of platform companies identified as key drivers of corporate transformation and potential allies in putting pressure on the companies. For example, ClimateVoice's 1in5 campaign appealed to the employees of major US-based platform companies. Driven by a cartoonish, superhero-themed video, the campaign called for tech company employees to be 'climate heroes' and to demand that their employers dedicate 20% of their political lobbying resources to advancing ambitious climate policies (ClimateVoice, 2024b). While the campaign did not engage in direct counter-accounting, in its communications ClimateVoice made use of a report by InfluenceMap, a philanthropically funded think tank, which found that the five major US-based platform companies directed only 4% of their lobbying resources to climate issues (InfluenceMap, 2021). The report advocated that platform companies adopt a more strategic and comprehensive approach to lobbying for climate change mitigation.

In addition to demanding greater climate policy engagement, environmental groups' digital advocacy has focused on platform companies' membership in many cross-sector industry groups that lobby against ambitious climate policies. This issue attracted widespread attention in the United States when corporate lobbying groups launched a series of activities in 2021 to derail the passage of the

Biden administration's climate legislation. The corruption watchdog group Accountable.US published a counter account naming dozens of the country's largest corporations as having helped block the legislation, including some of the major platform companies (Accountable.US, 2021). These revelations increased activist efforts to pressure platform companies to leave industry groups that lobby against climate policies. Recently, ClimateVoice has campaigned on this issue with the Escape the Chamber campaign, advocating for employees to push platform companies to leave the US Chamber of Commerce (Escapethechamber.org, n.d.), which is one of the country's most influential cross-sector industry groups.

Denouncing Business Collaboration With the Oil and Gas Industry

As a third focus area for digital advocacy, activists have intensified their criticism of business collaborations between platform companies and the fossil fuel industry. These campaigns underscore the ways in which platform companies provide essential services to fossil fuel companies, thereby facilitating the continued expansion of fossil fuel production. Growing activist interest in this issue mirrors recent developments whereby leading platform companies have strengthened their business ties with the oil and gas industry. Google, for example, established an oil, gas and energy division in 2018 and signed several deals with fossil fuel companies (Merchant, 2019). The following year, ExxonMobil and Microsoft announced the 'largest-ever oil and gas' partnership that would significantly boost oil production in the coming years (Romm, 2019). As critical accounts of these partnerships were published by tech news sites and progressive groups, environmental groups responded at the turn of the decade.

Some of the most visible digital advocacy in this area has been carried out, again, by Greenpeace. The focus on how platform companies promote the oil and gas industry also marked a notable shift towards a more confrontational approach in the group's campaigning. In 2020, it published a critical counter account that singled out Microsoft, Google and Amazon as 'climate hypocrites' and called them out for their affiliations with the oil and gas industry (Greenpeace, 2020b). Later the same year, the group published another report titled 'Oil in the cloud: How tech companies are helping big oil profit from climate destruction' (Greenpeace, 2020c). Both reports exemplify the naming and shaming tactics employed by environmental activist groups. These tactics aim to make strategic use of public shaming, as targeted companies may be concerned about potential damage to their reputation, loss of customers or difficulties in recruiting employees (Bloomfield, 2013).

Demanding Actions to Root Out Climate Disinformation and Online Harassment of Activists

The role of platform companies in shaping the contemporary information environment in which public debate over climate change takes place has become an increasing concern for environmental activists in recent years. An important driver behind this activism is the notion of climate mis- or disinformation, referring to the spread of false claims about climate change, whether inadvertently or by intent (Cook, 2019; Sethi, 2024). While awareness of how platform companies are involved in climate misinformation grew in the late 2010s (e.g., Shao et al., 2018), activists became more vocal about it at the beginning of the 2020s.

This issue was first addressed by the global online movement Avaaz, which launched digital advocacy campaigns targeting Google and Facebook for failing to root out climate misinformation on their platforms. In 2020, Avaaz released a counter account revealing that YouTube's recommendation algorithm was giving free promotion to videos containing climate misinformation, and that YouTube had created financial incentives through its advertising programme to spread climate misinformation (Avaaz, 2020). The following year, Avaaz's focus shifted to Facebook, following the report by InfluenceMap (2020) that demonstrated Facebook's deficiencies in blocking advertisements that spread climate disinformation. In its digital advocacy, Avaaz accused the company of offering a 'powerful tool' to spread misinformation about climate science and renewable energy (Avaaz, 2021a) and criticised Facebook for its failure to address the evolving tactics used to prevent effective climate action (Avaaz, 2021b).

Prominent ENGOs have since joined efforts to combat climate disinformation. In 2022, Friends of the Earth, Avaaz and Greenpeace USA engaged in collaborative counter-accounting with a report and scorecard that ranked prominent social media companies according to their policies to deal with disinformation and the transparency of those policies (Greenpeace, 2022). The same year, prominent ENGOs established the Climate Action Against Disinformation (CAAD) coalition, which includes anti-disinformation organisations. CAAD and its member organisations have published numerous counter accounts, building awareness of the threat platform companies and their current business model pose to climate action, including the online spread of climate misinformation (Climate Action Against Disinformation, 2024; Global Action Plan, 2022). ENGOs have also increasingly drawn attention to the hate speech and online harassment experienced by climate activists. For example, the activities of Global Witness have combined climate activism with the promotion of human rights, including digital rights. In its digital threats campaign, the organisation has produced counter accounts on how platform companies hinder climate action by enabling hate speech and harassment of climate experts (e.g., Global Witness, 2024a, 2024b).

Conclusion

Platform companies are unique corporate targets for activist groups due to the multifaceted nature of their operations and effects on climate change. In addition to being significant emitters due to their massive energy consumption, they have clout over the investment choices of utility companies, the climate advocacy of industry lobby groups and national and international climate policies. Moreover, their cloud and computing services are deeply integrated into the operation of almost all industries, including the energy sector. Finally, digital platforms also influence public opinion and shape lifestyles and consumption habits.

This chapter has explored how climate activism responds to this complex topic through issue selection. Examining climate groups' use of counter-accounting and digital advocacy aimed at enforcing platform companies' environmental accountability, we identified four distinct areas of accountability demands (see Table 8.2). It should be noted that our dataset was limited to the reports and campaigns of a small number of US-based and international ENGOs. The public prominence of these organisations notwithstanding, our findings should not be taken as representative of all climate activism directed at platform companies. Even so, the findings suggest that the accountability demands of ENGOs have expanded over time, corresponding with the growing economic, political and societal power of platform companies, as well as the increasingly critical scrutiny of their climate commitments. Whereas early activism focused on accelerating the shift to renewable energy, the 2020s have seen more emphasis on platform companies' climate policy advocacy, their ties with the oil and gas industry (e.g. Greenpeace, 2020c) and their responsibility for the spread of climate disinformation (e.g. Climate Action Against Disinformation, 2024). As awareness of the multiple roles of platform companies in impacting climate change has grown, an increasing number of international ENGOs and other groups have joined Greenpeace's efforts to hold these companies accountable. Our findings indicate that ENGOs have responded to the complexity of platform companies' involvement with climate change by extending the range of issues they address in digital advocacy. This reflects a more general trend in climate activism, which has become an increasingly divided field in terms of tactics and goal-setting (Hadden, 2015).

The flurry of digital advocacy aimed at enforcing the environmental and climate accountability of platform companies raises the question of its impact. While corporate responses to observed activism are beyond the scope of this chapter, there are some signs that ENGOs' activities have been impactful. First, the clean energy transition of US-based platform companies accelerated significantly over the 2010s, and part of that shift has been credited to the persistent campaigning of activist groups, particularly Greenpeace (Vaughan et al., 2023). This conclusion is in line with previous research suggesting that corporations, on average, do react to demands that originate from ENGOs, local communities and civil society at large (Delmas & Toffel, 2008; Grant & Vasi, 2017; Sharma & Henriques, 2005).

Second, there are certain indications that digital advocacy has had an impact on platform companies' approach to climate policy. While major platform companies continue to be members of influential lobby groups, such as the US Chamber of Commerce and the Business Roundtable, which have actively opposed ambitious climate policies, there are signs that companies have become increasingly vocal within those bodies to shift the direction of their climate policy advocacy (Hiar & Cama, 2024). Major US-based platform companies are also

Table 8.2. Environmental Non-governmental Organisations' Accountability Issues, Targets and Demands Related to Platform Companies and Climate Change.

Issue	Targets	Demands	Examples of ENGOs and Campaigns
Growing climate emissions	US-based companies (2009–2019) China-based companies (2017–)	Increase energy efficiency Shift to renewable energy sources	Greenpeace USA (Cool IT 2009–2013; Click Clean 2014–2019) Greenpeace East Asia (Clean Cloud 2020–)
Neglecting climate policy advocacy	US-based companies	Lobby ambitious climate policies Leave industry groups opposing climate action	Greenpeace USA (Click Clean 2014–2019) ClimateVoice (1in5 2020; Escape the Chamber 2024)
Facilitating oil and gas extraction	Amazon, Google, Microsoft	Stop collaboration with oil and gas companies	Greenpeace (ReThinkIT 2020)
Failure to root out climate disinformation and online harassment of activists	Facebook, X (Twitter), YouTube, TikTok, Pinterest	Prevent monetisation of climate disinformation Enhance online moderation	Avaaz (Detox the Algorithm 2020) Climate Action Against Disinformation (CAAD) Coalition (2022–)

actively pushing for greater transparency of the suppliers of data centres regarding their climate emissions (iMasons Climate Accord, 2024).

When it comes to business collaboration with oil and gas companies or actions to prevent climate disinformation, there are few indications that ENGO campaigns have been successful. Despite digital advocacy, major platform companies seem to continue with business as usual, reaping profits from advancing the fossil fuel industry (Biddle, 2023). Similarly, while platform companies have taken measures to root out climate disinformation in social media (Romero-Vicente, 2023), they have been less willing to change their business models. As an indication of the insufficiency of these measures, activists continue to denounce platforms for failing to prevent climate disinformation and for benefiting from advertisements that promote climate denial (Global Witness, 2024a, 2024b).

The mixed picture of ENGOs' success in putting pressure on platform companies is indicative of the limitations of activist campaigns aimed at driving corporate environmental accountability. As Davis et al. (2022) argue, campaigns targeting individual companies seldom generate the kind of impact that activists aim to achieve. While they may succeed in generating temporary accountability, companies continue doing business as usual once that attention fades (Newell, 2008), or merely engage in distractive discursive tactics to appear responsible (Frig et al., 2026). Even so, counter-accounts are likely to remain a key tool in activists' toolbox, and it is important to continue studying the relationship between counter-accounts and corporate accountability. In addition to pressuring corporations to be accountable, they facilitate social accountability by enabling public evaluation of corporate environmental activities (Dillard & Vinnari, 2017; Ojala et al., 2019).

What such general observations fail to address, however, are the specific challenges digital advocacy faces when it turns its focus towards platform companies. Relying on social media communications to generate support for their campaigns and bring visibility to their counter accounts, climate activists become dependent on the same companies that they seek to hold accountable. Over the past decade, the debate on platforms and their capacity to empower social movements, activism and civil society has become increasingly pessimistic (Kavada et al., 2023). Despite providing NGOs with useful tools, social media contributes to an overloaded information environment where it is increasingly difficult to gain attention (Guo & Saxton, 2018). Moreover, the increase of far-right groups, harassment of progressive activists and corporate disinformation campaigns, as well as the ubiquitous data collection and surveillance exercised by platforms, have discredited any optimism about digital platforms and social media as technologies of emancipation. However, Blum and Uldam (2024) show that environmental activists still perceive social media as a space that enables them to challenge corporations in ways that are not possible offline, even if the corporations they target have far more resources than they do to shape that space.

In this regard, Blum and Uldam (2024) have criticised climate activists for adhering to the commercial logics of platforms, such as using divisive tactics like shaming instead of confronting how platforms function as businesses. By contrast, our findings about the movement against climate disinformation suggest that climate activism actively challenges the business model of platform companies. The companies are pressured to address algorithm-controlled content visibility and to support activists in their fight against those who seek to manipulate and suppress climate action. By engaging in such campaigns, activists acknowledge the power of platform companies to shape the contemporary information landscape. In the coming years, that recognition may increasingly lead them to combine the fight against climate change with the struggle for digital and epistemic rights.

Acknowledgements

This work was supported by the Research Council of Finland [grant number 332751] and The Strategic Research Council (SRC) established within the Research Council of Finland [grant number 372226].

References

Accountable.US. (2021). *Over 50 major corporations that claim to be concerned about climate issues help lead the very trade groups fighting the 'biggest climate change bill ever*. https://www.accountable.us/wp-content/uploads/2021/10/2021-09-21-Corporate-Hypocrisy-On-Reconciliation-Climate.pdf

Avaaz. (2020). *Why is YouTube broadcasting climate misinformation to millions?*. https://avaazimages.avaaz.org/youtube_climate_misinformation_report.pdf

Avaaz. (2021a). *Facebook's climate of deception: How viral misinformation fuels the climate emergency*. https://secure.avaaz.org/campaign/en/facebook_climate_misinformation/

Avaaz. (2021b). *Meta-denial: How Facebook fails to keep up with the evolving tactics of today's climate misinformers*. https://secure.avaaz.org/campaign/en/climate_briefing_report/

Biddle, S. (2023). Google greenwashes a dirty partnership with climate-destroying Saudi Aramco. *The Intercept*. https://theintercept.com/2023/02/17/google-cloud-saudi-aramco-oil/

Bloomfield, M. J. (2013). Shame campaigns and environmental justice: Corporate shaming as activist strategy. *Environmental Politics*, *23*(2), 263–281. https://doi.org/10.1080/09644016.2013.821824

Blum, I., & Uldam, J. (2024). Faking, optimising and conceding to power: Social movement understandings of social media power. *New Media & Society*, *0*(0). https://doi.org/10.1177/14614448241266769

Climate Action Against Disinformation. (2024). *The AI threats to climate change*. https://caad.info/analysis/reports/the-ai-threats-to-climate-change/

ClimateVoice. (2024a). *About us*. https://climatevoice.org/about/

ClimateVoice. (2024b). *1in5 campaign*. https://climatevoice.org/past-campaigns/1in5/

Cook, J. (2019). Understanding and countering misinformation about climate change. In I. Chiluwa & S. Samoilenko (Eds.), *Handbook of research on deception, fake news, and misinformation online* (pp. 281–306). IGI-Global.

Coombs, W. T., & Holladay, S. J. (2015). How activists shape CSR: Insights from internet contagion and contingency theories. In A. Adi, G. Grigore, & D. Crowther (Eds.), *Corporate social responsibility in the digital age* (pp. 85–97). Emerald Publishing Limited. https://doi.org/10.1108/S2043-052320150000007007

Cox, R. J. (2013). *Environmental communication and the public sphere* (3rd ed.). Sage.

Damgaard, B., & Lewis, J. M. (2014). Accountability and citizen participation. In M. Bovens, R. E. Goodin, & T. Schillemans (Eds.), *The Oxford handbook of public accountability* (pp. 258–272). Oxford University Press.

Davis, G. F., King, B., & Soule, S. A. (2022). Do social movements improve corporate behavior? A discussion and research agenda. *Rutgers Business Review*, *7*(2), 139–152. https://rbr.business.rutgers.edu/article/do-social-movements-improve-corporate-behavior-discussion-and-research-agenda

Delmas, M. A., & Toffel, M. W. (2008). Organizational responses to environmental demands: Opening the black box. *Strategic Management Journal*, *29*(10), 1027–1055. https://doi.org/10.1002/smj.701

Denedo, M., Thomson, I., & Yonekura, A. (2017). International advocacy NGOs, counter accounting, accountability and engagement. *Accounting, Auditing & Accountability Journal*, *30*(6), 1309–1343. https://doi.org/10.1108/AAAJ-03-2016-2468

Dey, C. (2003). Corporate "silent" and "shadow" social accounts. *Social and Environmental Accountability Journal*, *23*(2), 6–9. https://doi.org/10.1080/0969160X.2003.9651696

Dillard, J., & Vinnari, E. (2017). A case study of critique: Critical perspectives on critical accounting. *Critical Perspectives on Accounting*, *43*, 88–109. https://doi.org/10.1016/j.cpa.2016.09.004

Doyle, J. (2007). Picturing the clima(c)tic: Greenpeace and the representational politics of climate change communication. *Science as Culture*, *16*(2), 129–150. https://doi.org/10.1080/09505430701368938

Dupire, M., Filbien, J.-Y., & M'Zali, B. (2022). Non-governmental organization (NGO) tweets: Do shareholders care? *Business & Society*, *61*(2), 419–456. https://doi.org/10.1177/0007650320985204

Escapethechamber.org. (n.d.). https://escapethechamber.org/

Frig, M., Dovbysh, O., & Laaksonen, S.-M. (2026). Greening the platforms or organisational bullshit? A typology of empty and misleading environmental communication by global platform companies. In S.-M. Laaksonen, M. Pantti, & O. Dovbysh (Eds.), *Platforms and the planet: Big tech, digital platforms and environmental responsibility* (pp. 167–182). Emerald Publishing Limited.

Global Action Plan. (2022). *New report explores Big Tech's dirty secret*. https://www.globalactionplan.org.uk/news/new-report-explores-big-tech-s-dirty-secret

Global Witness. (2024a). *Before US election, TikTok and Facebook fail to block harmful disinformation*. YouTube succeeds. https://www.globalwitness.org/en/campaigns/digital-threats/us-election-tiktok-and-facebook-fail-block-harmful-disinformation-youtube-succeeds/

Global Witness. (2024b). *Disinformation dollars: How Google earns money from and funds articles that promote climate denial*. https://www.globalwitness.org/en/campaigns/digital-threats/disinformation-dollars-how-google-earns-money-and-funds-articles-promote-climate-denial/

Grant, D., & Vasi, I. B. (2017). Civil society in an age of environmental accountability: How local environmental nongovernmental organizations reduce U.S. power plants' carbon dioxide emissions. *Social Forum*, *32*(1), 94–115. https://doi.org/10.1111/socf.12318

Greenpeace. (2010). *Make IT green: Cloud computing and its contribution to climate change*. https://www.greenpeace.org/international/publication/7099/make-it-green-cloud-computing-and-its-contribution-to-climate-change/

Greenpeace. (2011). *How dirty is your data?* https://www.greenpeace.org/international/publication/7196/how-dirty-is-your-data/

Greenpeace. (2017). *Clicking clean: Who is winning the race to build a green internet?* https://www.greenpeace.org/international/publication/6826/clicking-clean-2017/

Greenpeace. (2019). *Internet giants reject dominion's plan for more fossil fuels in Virginia*. https://www.greenpeace.org/usa/news/internet-giants-reject-dominions-plan-for-more-fossil-fuels-in-virginia/

Greenpeace. (2020a). *Greenpeace releases first-ever clean energy scorecard for China's tech industry*. https://www.greenpeace.org/eastasia/press/2846/greenpeace-releases-first-ever-clean-energy-scorecard-for-chinas-tech-industry/

Greenpeace. (2020b). *Microsoft, Google, Amazon – Who's the biggest climate hypocrite?* https://www.greenpeace.org/usa/microsoft-google-amazon-energy-oil-ai-climate-hypocrite/

Greenpeace. (2020c). *Oil in the cloud: How tech companies are helping big oil profit from climate destruction*. https://www.greenpeace.org/usa/reports/oil-in-the-cloud/

Greenpeace. (2022). *In the dark: How social media companies' climate disinformation problem is hidden from the public*. https://www.greenpeace.org/usa/reports/in-the-dark/

Greenpeace. (2024). *Baidu lags behind Alibaba and Tencent in Greenpeace's latest China tech ranking*. https://www.greenpeace.org/eastasia/press/8575/baidu-lags-behind-alibaba-and-tencent-in-greenpeaces-latest-china-tech-ranking/

Greenpeace USA. (2019). *Ask Amazon: Why So much dirty energy to power the cloud?* https://www.youtube.com/watch?v=K7YXZeRT5bQ

Guo, C., & Saxton, G. D. (2018). Speaking and being heard: How nonprofit advocacy organizations gain attention on social media. *Nonprofit and Voluntary Sector Quarterly*, *47*(1), 5–26. https://doi.org/10.1177/0899764017713724

Hadden, J. (2015). *Networks in contention: The divisive politics of climate change*. Cambridge University Press.

Hiar, C., & Cama, T. (2024). *Climate strife divides US chamber. Will departures follow?* E&E News. https://www.eenews.net/articles/climate-strife-divides-us-chamber-will-departures-follow/

iMasons Climate Accord. (2024). *AWS, digital realty, Google, Meta, Microsoft and Schneider electric call for supplier action to help decarbonize digital infrastructure*. https://climateaccord.org/press-releases/call-for-supplier-action-to-help-decarbonize-digital-infrastructure/

InfluenceMap. (2020). *Climate change and digital advertising: Climate science disinformation in Facebook advertising*. https://influencemap.org/report/Climate-Change-and-Digital-Advertising-86222daed29c6f49ab2da76b0df15f76#1

InfluenceMap. (2021). *Big Tech and climate policy*. https://influencemap.org/EN/report/Big-Tech-and-Climate-Policy-afb476c56f217ea0ab351d79096df04a

Johansson, H., & Scaramuzzino, G. (2019). The logics of digital advocacy: Between acts of political influence and presence. *New Media & Society*, *21*(7), 1528–1545. https://doi.org/10.1177/1461444818822488

Karpf, D. (2020). After clicktivism. In W. H. Dutton (Ed.), *A research agenda for digital politics* (pp. 123–132). Edward Elgar.

Katz-Kimchi, M., & Manosevitch, I. (2015). Mobilizing Facebook users against Facebook's energy policy: The case of greenpeace unfriend coal campaign. *Environmental Communication*, *9*(2), 248–267. https://doi.org/10.1080/17524032.2014.993413

Kavada, A., Askanius, T., Kaun, A., Mattoni, A., & Uldam, J. (2023). *Tying up goliath: Activist strategies for confronting and harnessing digital power*. Transnational Institute. https://www.tni.org/en/article/tying-up-goliath

King, B. G. (2008). A political mediation model of corporate response to social movement activism. *Administrative Science Quarterly*, *53*(3), 395–421. https://doi.org/10.2189/asqu.53.3.395

King, B. G., & McDonnell, M.-H. (2015). Good firms, good targets: The relationship between corporate social responsibility, reputation, and activist targeting. In A. Lim & K. Tsutsui (Eds.), *Corporate social responsibility in a globalizing world* (pp. 430–454). Cambridge University Press. https://papers.ssrn.com/sol3/papers.cfm?abstract_id=2079227

Laaksonen, S.-M., & Premsankar, G. (2026). The challenging task of estimating the carbon footprint of digital services. In S.-M. Laaksonen, M. Pantti, & O. Dovbysh (Eds.), *Platforms and the planet: Big tech, digital platforms and environmental responsibility* (pp. 37–52). Emerald Publishing Limited.

Maktoufi, R., O'Connor, A., & Shumate, M. (2020). Does the CSR message matter? Untangling the relationship between corporate–nonprofit partnerships, created fit messages, and activist evaluations. *Management Communication Quarterly*, *34*(2), 188–212. https://doi.org/10.1177/0893318919897059

McDonnell, M.-H., King, B. G., & Soule, S. A. (2015). A dynamic process model of private politics activist targeting and corporate receptivity to social challenges. *American Sociological Review*, *80*(3), 654–678. https://doi.org/10.1177/0003122415581335

Merchant, B. (2019). *How Google, Microsoft, and Big Tech are automating the climate crisis*. https://gizmodo.com/how-google-microsoft-and-big-tech-are-automating-the-1832790799

Moerman, L., & van der Laan, S. (2015). Exploring shadow accountability: The case of James Hardie and asbestos. *Social and Environmental Accountability Journal*, *35*(1), 32–48. https://doi.org/10.1080/0969160X.2015.1007464

Moore, M. H. (2014). Accountability, legitimacy, and the court of public opinion. In M. Bovens, R. E. Goodin, & T. Schillemans (Eds.), *The Oxford handbook of public accountability* (pp. 632–646). Oxford University Press.

Newell, P. (2008). Civil society, corporate accountability and the politics of climate change. *Global Environmental Politics*, *8*(3), 122–153. https://doi.org/10.1162/glep.2008.8.3.122

Ojala, M., Pantti, M., & Laaksonen, S.-M. (2019). Networked publics as agents of accountability: Online interactions between citizens, the media and immigration officials during the European refugee crisis. *New Media & Society*, *21*(2), 279–297. https://doi.org/10.1177/1461444818794592

Rodrigue, M. (2014). Contrasting realities: Corporate environmental disclosure and stakeholder-released information. *Accounting, Auditing & Accountability Journal, 27*(1), 119–149. https://doi.org/10.1108/AAAJ-04-2013-1305

Romero-Vicente, A. (2023). Platforms' policies on climate change misinformation. *EU Disinfo Lab.* https://www.disinfo.eu/publications/platforms-policies-on-climate-change-misinformation/

Romm, J. (2019). *Amazon, Google, and Microsoft are quietly helping big oil destroy the climate.* https://archive.thinkprogress.org/amazon-google-microsoft-oil-climate-b8b1dd15f115/

Saxton, G. D., & Waters, R. D. (2014). What do stakeholders like on Facebook? Examining public reactions to nonprofit organizations' informational, promotional, and community-building messages. *Journal of Public Relations Research, 26*(3), 280–299. https://doi.org/10.1080/1062726X.2014.908721

Sethi, P. (2024). *What are climate misinformation and disinformation and what is their impact?* Grantham Research Institute. https://www.lse.ac.uk/granthaminstitute/explainers/what-are-climate-misinformation-and-disinformation/

Shao, C., Hui, P. M., Wang, L., Jiang, X., Flammini, A., Menczer, F., & Ciampaglia, G. L. (2018). Anatomy of an online misinformation network. *PLoS One, 13*(4), e0196087. https://doi.org/10.1371/journal.pone.0196087

Sharma, S., & Henriques, I. (2005). Stakeholder influences on sustainability practices in the Canadian forest products industry. *Strategic Management Journal, 26*(2), 159–180. https://doi.org/10.1002/smj.439

She, C. (2022). Social media dissemination of counter accounts and stakeholder support – Evidence from Greenpeace's 'Save the Arctic' campaign on Facebook. *Accounting Forum, 47*(3), 390–415. https://doi.org/10.1080/01559982.2021.2019524

Thomson, I., Dey, C., & Russell, S. (2015). Activism, arenas and environmental reporting in conflicts over tobacco control. *Accounting, Auditing & Accountability Journal, 28*(5), 809–845. https://doi.org/10.1108/AAAJ-08-2013-1439

Vaughan, H., Pasek, A., Silcox, N. R., & Starosielski, N. (2023). ICT environmentalism and the sustainability game. *Journal of Language and Politics, 22*(5), 640–660. https://doi.org/10.1075/jlp.22125.vau

Vigren, M., Karppi, T., & Pyyhtinen, O. (2026). Calculating, reducing, compensating and disconnecting: Summons of digital excess. In S.-M. Laaksonen, M. Pantti, & O. Dovbysh (Eds.), *Platforms and the planet: Big tech, digital platforms and environmental responsibility* (pp. 67–84). Emerald Publishing Limited.

Vinnari, E., & Laine, M. (2017). The moral mechanism of counter accounts: The case of industrial animal production. *Accounting, Organizations and Society, 57*, 1–17. https://doi.org/10.1016/j.aos.2017.01.002

Chapter 9

Challenging Plastic Polluters: Public Shaming Campaigns on Plastic Pollution on Social Media Platforms

Steph Hill

University of Leicester, UK

Abstract

This chapter examines the Break Free from Plastic (BFFP) coalition's 'brand audit' campaigns, which specifically focus on corporate accountability and public narratives around responsible handling of plastic waste. By analysing the use of social media platforms in a global environmental campaign, the chapter focuses on how activist groups channel social accountability efforts through the logic of platforms in the media ecosystem. The chapter draws on BFFP's own documentation of the campaigns and instructions to activists as well as analysis of associated blog posts, strategy documents and social media toolkits. It examines how the coalition understands media strategy for the brand audit and corporate accountability campaigns in a complex media environment that enables the networked campaign while raising questions about the power relationship between social media platforms, businesses and activist groups in these efforts. This chapter asks how does the brand audit campaign produce pressure on companies through social media practices, and how do companies respond? I argue that the brand audit campaigns demonstrate an evolution in the strategy of pressuring businesses over complex environmental issues, part of a trend in activism that is itself indicative of how platforms inform the conditions for environmental activism and discussion.

Keywords: Social media platforms; plastic pollution; brands; activism; environmental campaigning; corporate accountability

Platforms and the Planet, 153–165

doi:10.1108/978-1-83662-172-020261010

Introduction

The #BreakFreeFromPlastic (BFFP) campaign announces its social media orientation from the very first hashtag. The organisation itself is a network of thousands of environmental organisations that cooperate on documenting and organising against plastic pollution. Among their most well-known campaigns is an annual 'brand audit' (piloted in 2017) in which the group catalogues the visible branding on waste found in clean ups around the world to create a top 10 list of the world's most prolific plastic polluters. Those identified are widely named and shamed in annual 'Branded' reports, social media posts and media reporting of the audit's results. This chapter examines the BFFP campaign to understand how the platform companies inform the conditions for environmental activism.

By analysing the use of social media platforms in a global environmental campaign, the chapter focuses on how activist groups channel social accountability efforts (Bovens, 2007) through the logic of platforms in a hybrid media ecosystem. Social media platforms provide social movement actors with cheap, fast and networked means of communication. However, at the same time, they raise concerns about their empowering potential, especially concerning social movement actors' possibilities to harness the power of social media for holding businesses accountable for not acting in socially responsible ways.

The brand audit campaign combines assessment of the brands represented in plastic rubbish collected on sites such as beaches with reports, press releases and public pressure activities directed at offending companies, including engaging the public in social media campaigns directed at the worst offenders. BFFP credits the campaign with changing media narratives around plastic pollution. In particular, they claim that the campaign has been pivotal in shifting discussion of responsibility for plastic pollution from downstream issues, such as individual recycling rates and beach clean ups, towards upstream issues such as production, design and producer responsibility. BFFP calls what they are doing 'corporate accountability', a term that is not widely used in academic research, where similar activity is more likely to be called 'social accountability' (Bovens, 2007) or addressed as part of the research on consumer activism (Lekakis, 2022). Organisation and management scholars once suggested that accountability campaigns powered by social media, as the BFFP campaign is, would help to end corporate greenwashing by enabling the public to effectively monitor corporate activity and pressure hypocritical or underperforming businesses (Lyon & Montgomery, 2013). Ten years later, those same scholars are the first to recognise that the optimistic scenarios have failed to materialise (Montgomery et al., 2023). Instead, greenwashing concerns have proliferated and become the focus of regulatory attention by advertising standards bodies, as well as national and international governments.

The chapter draws on BFFP's own documentation of the campaigns and instructions to activists as well as analysis of associated blog posts, strategy documents and social media toolkits. It examines how the coalition understands media strategy for the brand audit and corporate accountability campaigns in a complex media environment that enables the networked campaign itself while raising questions about the power relationship between social media platforms, businesses and activists. This chapter compares the brand audit campaign's goals and claims of success with how the consumer goods companies that it targets engage with the issue of plastic pollution in forums such as the United Nations Environment Assembly (UNEA). It asks how does the brand audit campaign produce pressure on companies using social media, and how do companies respond? Drawing on theories of activist media and institutionalist frameworks, I argue that the brand audit campaigns demonstrate an evolution in the strategy of pressuring well-known businesses over complex environmental issues, part of a trend in activism that is indicative of how social media platforms inform the conditions for environmental activism and discussion.

Platforms and Activist Media Methods

Corporate social media platforms have become main channels for environmental activist groups to attract visibility and support for their campaigns' targets (see also Chapter 10 in this volume). Social media platforms are theorised as an aid to groups trying to hold organisations accountable for environmental problems. This optimism is sometimes directly tied to social media as an enabling force for corporate accountability campaigns (Lyon & Montgomery, 2013).

Activists themselves have sophisticated understandings of how social media platforms allow them to amplify their presence and game algorithms to 'optimise' their communication, as well as how the economics of platform algorithms can disadvantage them (Blum & Uldam, 2024). The power asymmetries created by social media platforms are evident in how the social media platforms are exploiting the user data created by the activist activity on platforms: the more visibility environmental groups are creating, the more profit they create for the platform company (Kavada et al., 2023). Crucially, in Blum and Uldam's (2024) study, activist groups see no alternative to using and adapting to the logics of the largest social media platforms, despite their awareness of their commercial orientation. The economy of social media platforms ties the functioning of the platform closely to their advertisers' interests and creates an environment where brands' participation in social conversation is normalised (Van Dijck, 2013). Brands have always been defined through their interactions with consumers. However, digital technologies, including social media platforms, have enabled a 'calculative' turn in this relationship where brands 'attune themselves to the identities and cultural lives of their consumers' by drawing on data collected through online surveillance (Carah, 2017). The management of data and its use in controlling how content is delivered is crucial to the functioning of 'algorithmic media', where management of visibility and invisibility is a vital concern for organisations of all kinds (Bucher, 2012; Flyverbom et al., 2016). The importance of strategic visibility on social media platforms has facilitated adaptations and amplifications of activist tactics, including activities such as mobilising supporters and 'jamming' pop cultural objects created by companies (Lievrouw, 2023).

Social media can play different roles in activism, and there are reasons to differentiate between internet-based and internet-enabled social movements (van Laer & van Aelst, 2010), as well as pay attention to the specifics of the activism that platform media enables (Lievrouw, 2023). Examining the BFFP campaign provides a case study of how activist groups use platform vernaculars, the social practices linked to particular platforms and their affordances (Gibbs et al., 2015). In particular, the brand audit campaign incorporates the hashtag activism associated with Twitter and image-sharing associated with Instagram, along with direct communication with brands on platforms, to enact forms of social accountability on their corporate targets. Insofar as the campaign is focused on recognisable consumer brands, the brand audit campaign has some elements of consumer activism (Lekakis, 2022; Micheletti et al., 2011). However, analysis of the campaign itself raises questions over whether the purpose of the campaign is to change the behaviour and practices of large brands or to engage with broader narratives of responsibility for plastic pollution.

Accountability, Organisational Responsibility and the Case of #BreakFreeFromPlastic

Pressure from external groups represents a particular idea of how accountability happens, one that is constantly debated. Bovens (2007) wrote that 'It remains an empirical question to what extent these groups and panels [interest groups, NGOs, and stakeholders] already are full accountability mechanisms, because the possibility of judgment and sanctioning often are lacking' (p. 457). So, one question that is important to examining the BFFP campaign is how accountability is understood by the campaign, and whether there is evidence that it is taken seriously by its targets. In other words, is public shaming on social media working as an accountability mechanism?

To parse questions of accountability, this chapter draws on institutionalist frameworks that are attentive to interactions between institutions and their wider context (Bannerman & Haggart, 2015; Powell & DiMaggio, 2012). Institutionalist frameworks have been used to explain differences in companies' behaviour within the oil and gas industry engagement with environmental issues as 'bottom-up processes of interpretation and contestation' (Pulver, 2007, p. 49). In some cases, scholars found that 'particular combinations of structural power and cultural vulnerability almost guaranteed that firms would become major targets of activism' (Bartley & Child, 2014, p. 655). Cultural vulnerabilities included social conditions such as branding, reputations within their industry and perceived commitment to social responsibility, which combined to make some firms more 'shamable' by social movements than others, a process sometimes dubbed the 'social production of targets' (Bartley & Child, 2014). The socially produced 'shame-ability' of targets is a useful lens for understanding the BFFP campaign, particularly as social media platforms add another dimension to the 'cultural vulnerability' that Bartley and Child identified. At the same time, it is important to consider whether and how these institutionalist frames account for power, economic incentives and conflicts of interest in policy environments.

Historically, researchers find that corporate actors used their power to advocate for solutions, such as recycling, that allow businesses to continue current practices and shift responsibility to individuals and governments (CCI, 2024; MacBride, 2013). For example, advocacy groups such as Keep America Beautiful represented packaging and product companies and focused attention on litter rather than production and incineration over source reduction of waste materials (Hess, 2007). More recently, scholars have identified potential deception and negative engagement from businesses in regard to plastic. For instance, Mah (2021) identifies corporate initiatives, such as the Alliance to End Plastic Waste, as greenwashing enterprises that cloak the ongoing production of problematic plastics with the appearance of corporate social responsibility. However, some scholars also see businesses as receptive to pressure from social movements. Lounsbury et al. (2003) identify the social structures and meanings produced by non-profit recyclers as a 'field frame' that eventually transformed the practices of the for-profit recycling industry in the United States.

We can ask whether these accountability measures also legitimise the corporations that they target as governance intermediaries, as Braun et al. (2019) ask about Sleeping Giants' campaign against online advertisers. In the case of Sleeping Giants, a campaign that targeted online advertisers as intermediaries with the power to restrict extremist content, the authors suggest that the campaign's effectiveness simultaneously 'bolsters the system of advertiser-support, and the legitimacy of corporations censoring media content' (Braun et al., 2019, p. 73). They note the 'greater symmetry between users and brands on social media' as a condition of Sleeping Giants' success (p. 72).

To analyse the #breakfreefromplastic brand audit campaign, I patterned this analysis after case study research (Yin, 2017) and studies that create 'historical narratives' of organisations using publicly available media (Brulle, 2022). Materials reviewed included the 5 years of brand audit reports (2018-2023), the social media toolkits published (2020, 2021) and organisational documents such as the 'Break Free From Plastic Communication Cookbook – Asia Edition' (2021). I also reviewed all 103 posts on the BFFP website tagged as 'corporate accountability'.

To assess how companies responded to this pressure, I reviewed the submissions of all the companies targeted by the Brand Audit to the global plastic treaty process, whether submitted independently or as part of larger associations, and all blogs and policy documents published by the Business Coalition for a Global Plastic Treaty (a group that includes 8/10 of the BFFP's list of the top plastic polluters found in brand audits), the Consumer Goods Forum or organising groups such as the Ellen MacArthur Foundation and Global Plastic Action Partnerships. This document analysis was supplemented by observation at events related to the brand audits, corporate accountability in plastic pollution and negotiations at the UNEA, such as the meetings of the Intergovernmental Negotiating Committee on Plastic Pollution.

Given the large corpus of documents and the research questions of this chapter, I focused initially on the five annual brand audit reports to understand the campaign's primary formal goals and framing. This analysis was followed by close examination of public responses from the targeted companies, as well as a review of their submissions to the global plastics treaty process to understand whether and how they responded to this public pressure. This was a qualitative process and focuses on what I believe to be key themes and strategies that underline the close connections between platform structures and economies and environmental campaigning and narratives.

The Brand Audit Campaign and its Effects

The #BreakFreeFromPlastic campaign's inclusion of a hashtag in its name highlights its connection to social media activism. The campaign's communication strategies are laid out in a 'cookbook' created for the group by Dancing Fox Communications, a consultancy that works closely with environmental groups. In the communications cookbook, Dancing Fox encourages the campaign to practice 'invitational activism' that turns 'heroes into recruiters' (Dancing Fox, 2021, p. 10) by 'amplify[ing]' the actions of ordinary people and choosing strategic messengers. The cookbook suggests that the campaign must create content that is 'so Cute or Outrageous or Offensive or Beautiful' [capitalisation in the original] that people feel the need to discuss it and share it, allowing the campaign to 'punch through the noise' of social media platforms like TikTok, We-Chat and Instagram. The 2017 video of a sea turtle with a plastic straw up its nose (Rosch, 2024) is presented as a good example of this kind of content. In addition, the cookbook argues that the campaign should create a 'jam-culture' that is 'designed to disrupt mainstream cultural opinions and behaviour by subverting the most powerful products of mainstream media' (p. 34). This approach, like 'culture jamming' (Lievrouw, 2023) and 'brandalism' (Lekakis, 2017), works by 'piggy-backing on something that already has high levels of traction within culture' (Dancing Fox, p. 35), such as recognisable brand colours and fonts.

A review of the campaign's online coverage, using the hashtags #breakfreefromplastic and #BrandAudit[year], demonstrates how some of these tactics are put into practice, revealing images of animals suffering from ingesting plastic, messages encouraging individuals to tag grocery stores and aligned organisations announcing the publication of the campaign's 'Branded' reports. Instagram searches reveal a grid of identical social media-friendly graphics of the top 10 list of 'worst plastic polluters'. Many of these images were created and shared by the BFFP campaign in its social media toolkits – packages of images and sample posts designed to be shared on social media. Others seem to be taken as screenshots directly from the campaign reports and shared as images.

The BFFP social media toolkits provide materials that activists can use to 'encourage YOUR community to join #BrandAudit'. Sample tweets include 'Hey @cocacola, @pepsico, @nestle clean-ups are not enough! Let's hold companies accountable with #BrandAudit2021' and 'It's time corporations own up to the truth. Join #BrandAudit2021 and let's hold corporations accountable for their plastic waste'. Sample images include Instagram-sized images with text, including '#1 WORST PLASTIC POLLUTER' accompanied by the Coca-Cola logo and several of its most well-known properties. Similar images are available for other front-running companies, such as PepsiCo and Nestle. Graphics are available for each of the top 10 brands identified in the brand audits, as well as the full top 10 list, and the toolkit includes images and videos that can be used in Instagram stories and posts. These platform-focused materials are one facet of the campaign, but examining the campaign's website, timing of posts and definitions of success reveals a more complex picture of its relationship to media visibility.

BFFP Is a Hybrid, Media-Focused Campaign

One of the first items visible on the BFFP's Brand Audit website, just under 'what is a brand audit', is a banner reading 'as seen on' and showcasing media coverage in the BBC, *The Guardian, Le Monde* and *Last Week Tonight*. The page links to sample articles that showcase how BFFP would prefer to see its findings presented. These samples appear in major legacy media outlets and emphasise major brands as key polluters, with headlines such as 'Coca-Cola's 100 Billion Bottle Problem' (Mulholland & Stern, 2021). Slightly lower on the page is a button inviting journalists to 'Feature Brand Audit' and links again to these high-profile media features of the campaign and its results, inviting journalists and the press to cover the brand audit in a similar fashion. The campaign launches its results online but demonstrates how much it values the coverage from legacy sources of media in the prominence they are given on its site and how it describes the campaign's successes. Examining the posts tagged as part of BFFP's corporate accountability campaign reveals posts that are timed to overlap with newsworthy moments, such as announced failures to meet targets set under voluntary efforts such as the Ellen MacArthur Foundation's Global Commitment, new studies on plastic pollution, and events such as recycling week and World Cleanup Day. These details suggest a campaign that targets legacy press coverage and media attention in tandem with platform media strategies, a view the campaign seems to confirm in how it defines success.

The campaign defines success in terms of changing media narratives. In the BFFP's 2020 report, they describe the success of the campaign in terms of public narrative, as demonstrated by Google search: 'Five years ago, an internet search for 'top plastic polluters' yielded a list of countries due to the now-retracted Stemming the Tide report. Today, this same search results in a list of the top plastic polluting corporations as a direct result of our annual global brand audits' (BFFP, 2020, p. 25). The focus on brands in the audits is intended as a replacement for narratives of individual responsibility and of plastic pollution as primarily a problem for developing countries. In other words, the campaign is an effort at agenda-setting for media narratives about plastic pollution that explicitly incorporates platform dynamics such as prominence in search to achieve its aims.

The insistence on a shift in responsibility for plastic pollution is consistent throughout the branded reports. Responsibility for plastic pollution is described as '...not with individual "litterbugs", but with corporate polluters who must adopt sustainable solutions and systems to stop the crisis' (BFFP, 2018, p. 52019, 152019, 6). The brand audit itself is argued to challenge or restrict the ability of corporate actors to 'frame the issue as one of only consumer responsibility' (2019, p. 15) when '...it is mainly companies based in Europe and the United States that are producing massive quantities of single-use plastic packaging and spreading it all over the world' (2019, p. 6). The campaign claims it is 'challenging industry narratives about who is responsible for the plastics crisis and how to solve it' (2020, p. 15). The purpose of the campaign is clear: re-shape the narrative of responsibility for plastic pollution by focusing attention on branded plastic litter.

Petrochemical companies are not targeted under the BFFP corporate accountability campaigns and brand audits, even though the oil and gas industry is certainly run by corporations. Instead, the organisation has a separate campaign for the petrochemical industry. The materials are overwhelmingly focused on 'the brands' in the global fast-moving consumer goods category. The methods of the audit rank polluters based on the most plastic found in the most countries – a choice that screens out the prevalence of 'regional or national brands' (p. 16). This is consistent with consumer goods companies being more responsive to criticism, having a wider public profile and being more accessible for activists.

In the reports of the brand audit campaigns, the justifications for focusing on the brands appear in various forms. While the reports mention other sources and forms of plastic pollution, such as communities that are located near plastic production plants, the ability to name and shame 'recognizable brands' (BFFP, 2018, p. 112021, 3) is critical to the campaign both online, through tagging the companies directly, and through media coverage. Other sources of plastic pollution are more difficult to attribute: 'the nurdles found on beaches don't have recognizable brand logos printed on them' (p. 11). The campaign singles out the 43% of found waste that is 'marked with a clear consumer brand' (2021, p. 3) because 'these corporations have a name and address' (p. 8). Outside of the audit reports, the same strategy of corporate accountability is maintained. For instance, at the time of this research (spring 2024), ad-style boxes appeared on the website's right-hand side quoting a study (Cowger et al., 2024) linking most of the plastic pollution to '56 identifiable companies'. Giving prominence to this study, which uses brand audit data, reinforces a central place for companies in responsibility for plastic pollution and encourages sharing of the study as newsworthy identification of responsibility for plastic pollution, though it reinforces attention to the most visible, identifiable and easily contacted offenders.

Corporate Communication

Does this media-narrative focused campaign succeed? While it is difficult to absolutely prove causation, there are some reasons to think that the targets of BFFP's brand audit campaign have received the message. Looking specifically at the brand audit's primary target, Coca-Cola's public statements on waste management demonstrate a shift over time that mirrors the shift in public discussion of plastic waste more generally. In 2018, Coca-Cola launched their 'World Without Waste' initiative, accompanied by a flurry of new partnerships and investments in recycling, clean up and partnerships. In 2020, Coca-Cola publicly defended their investments in plastic because 'customers still want them' (Joshi, 2020). This comment was widely circulated by media outlets, including *Vice*, the BBC, the *Verge* and *Forbes*, alongside the framing of the company as 'one of the biggest producers of plastic waste' and references to BFFP's brand audits (Thomas, 2020). The attention to the comment is itself a testament to BFFP's ability to frame the company as responsible for plastic pollution. By 2022, Coca-Cola announced new targets for products sold in refillable and reusable containers. It had also become part of a new coalition engaged with global plastics policy, the Business Coalition for a Global Plastics Treaty.

There is some evidence that the target industries of the brand audit campaign have developed policy positions that incorporate more aspects of reduction, reuse and re-design over time, along with and occasionally instead of downstream measures, such as recyclability and investment in waste management. In the negotiations for a global instrument on plastic pollution at the UNEA, representatives of the beverage industries, as well as the Business Coalition regularly contribute to the discussions. They advocate for extended producer responsibility, elimination or problematic plastics, and circular design of packaging and have developed to in-depth policy positions on restrictions, re-use, product design, extended producer responsibility and waste management. These positions are generally consistent with those of the High Ambition Coalition countries, as well as many advocacy groups. For example, the submission of the International Council of Beverage Associations to the second meeting of the Intergovernmental Negotiating Committee (2023) describing the principles that should animate any eventual treaty includes focus on the waste hierarchy (in which reduction comes before recycling), as well as social rights, such as support for informal waste pickers.

The examples above indicate a shift in how accountability for plastic pollution is discussed by the consumer-facing brands targeted by the brand audit campaign and how those companies and their representatives have moved towards their critics in environmental campaigning organisations over time.

Discussion and Conclusion

The brand audit campaign can seem removed from social media platforms' direct impacts on sustainability. However, the campaign and its targets are both enmeshed in the structures and media logics of social media platforms. Like other activist movements that began online, the audit relies on communication technology that enables the dissemination of videos that instruct global activists on how to conduct a brand audit (hosted on YouTube) and recruit (activists are encouraged to recruit through Instagram, Facebook and X/Twitter). While it is easy to forget how difficult these steps were previously, the brand audit relies on the key aspects of these platforms, including their recombinant nature, cultural ubiquity and interactivity (Lievrouw, 2023) in order to achieve its scale and recognition.

The brand audit campaign is networked from data collection to the dissemination of results. As an activist campaign, it enables 'connective action' by individuals not necessarily affiliated or identified with the core organising group (Bennett & Segerberg, 2012). As the campaign suggests, the ability to network global efforts is critical to what the campaign hopes to achieve: 'By gathering data on plastic waste collected at community cleanups around the world, brand audits allow members and supporters of the Break Free From Plastic movement to challenge the plastic industry, hold plastic polluting companies accountable, and build a global community for change' (2022, p. 7).

For the campaign's targets, the structure of social media platforms makes it easier for those brands to become visible targets for criticism. BFFP's brand audit takes advantage of these affordances and the mutual reinforcing of platform and legacy media narratives to challenge and crowd out other narratives of who is responsible for plastic pollution with a focus on consumer brands. This complementary strategy of utilising platform and legacy media echoes other works on online media agenda setting (Langer & Gruber, 2021) and indicates some of the ways that platforms inform the conditions for environmental activism and discussion. While this campaign has corporate targets, it is not 'marketplace-oriented'. It does not use boycotts or discourage consumers from using products made by the companies that are targeted. Rather, the brand-audit continues a shift towards 'media-oriented' campaigns that target companies in an effort to secure media attention (Friedman, 1999). The purpose of the campaign is not necessarily limited to, or even primarily intended to, change corporate behaviours around plastic production and use. Rather, it targets recognisable consumer brands as a means to define the boundaries of acceptable behaviour (de Bakker, 2015; den Hond & de Bakker, 2007) and change the framing of the global governance of plastic materials and plastic waste. In this case, social accountability uses the institutional logics of branded corporations and their visibility in culture and media to change the way accountability for plastic pollution is attributed.

In the BFFP campaign, we see a hybrid media campaign, where platform vernaculars, such as shareable images, tagged posts and search engine results, coincide and reinforce other agenda-setting elements. In this campaign, the conditions of the social media platforms are part of the 'social construction of targets' for activism (Bartley & Child, 2014) and inform the focus to hold these corporations accountable in *this* way. For this campaign, the brands are identifiable, accessible and responsive and therefore 'shamable' in ways that other groups, including much larger oil and gas companies, who see the production of plastic as a growth area, are not. They are branded, which enables activists to take advantage of millions of dollars of investment in name and brand recognition and use that recognition to draw attention to plastic pollution and its governance using culture-jamming tactics. They are online, which enables activists to take advantage of their visibility to tag them in posts and encourage the public to target them directly (Kavada et al., 2023). However, visibility online works both ways. Businesses with ample economic resources have opportunities to address social media pressure by paying for reach on social media platforms and buying space to disseminate their own narratives (Blum & Uldam, 2024).

The brand audit's corporate accountability campaign protests the role of these corporate actors as creators of global plastic pollution, but it also positions these companies as necessary leadership. The brand audit focuses on companies that have the kinds of investment in branding and public communication that require them to respond to public criticism. It is these investments that make the companies accountable. Surveillance of online criticism also facilitates corporate strategies that anticipate and respond to critique (Uldam, 2016). Businesses can then act to safeguard their legitimacy using time-tested strategies, such as self-regulation and corporate social responsibility activities, or more tailored responses to online accounts. When BFFP pushes these companies to demonstrate leadership, it also legitimises their efforts and reinforces their visibility as important actors in environmental governance.

References

Bannerman, S., & Haggart, B. (2015). Historical institutionalism in communication studies. *Communication Theory*, *25*(1), 1–22. https://doi.org/10.1111/comt.12051

Bartley, T., & Child, C. (2014). Shaming the corporation: The social production of targets and the anti-sweatshop movement. *American Sociological Review*, *79*(4), 653–679. https://doi.org/10.1177/0003122414540653

Bennett, W. L., & Segerberg, A. (2012). The logic of connective action: Digital media and the personalization of contentious politics. *Information, Communication & Society*, *15*(5), 739–768. https://doi.org/10.1080/1369118X.2012.670661

BFFP. (2018). *Branded: In search of the world's top corporate plastic polluters.* https://brandaudit.breakfreefromplastic.org/brand-audit-2018/

BFFP. (2019). *Branded: The brand audit report 2019.* https://brandaudit.breakfreefromplastic.org/brand-audit-2019/

BFFP. (2020). *Branded: The brand audit report 2020.* https://brandaudit.breakfreefromplastic.org/brand-audit-2020/

BFFP. (2021). *Branded: The brand audit report 2021.* https://brandaudit.breakfreefromplastic.org/brand-audit-2021/

Blum, I., & Uldam, J. (2024). Faking, optimising and conceding to power: Social movement understandings of social media power. *New Media & Society*, *0*(0). https://doi.org/10.1177/14614448241266769

Bovens, M. (2007). Analysing and assessing accountability: A conceptual framework. *European Law Journal*, *13*(4), 447–468. https://doi.org/10.1111/j.1468-0386.2007.00378.x

Braun, J. A., Coakley, J. D., & West, E. (2019). Activism, advertising, and far-right media: The case of Sleeping Giants. *Media and Communication*, *7*(4), 68–79. https://doi.org/10.17645/mac.v7i4.2280

Brulle, R. J. (2022). Advocating inaction: A historical analysis of the Global Climate Coalition. *Environmental Politics*, *32*(2), 185–206. https://doi.org/10.1080/09644016.2022.2058815

Bucher, T. (2012). Want to be on the top? Algorithmic power and the threat of invisibility on Facebook. *New Media & Society*, *14*(7), 1164–1180. https://doi.org/10.1177/1461444812440159

Carah, N. (2017). Algorithmic brands: A decade of brand experiments with mobile and social media. *New Media & Society*, *19*(3), 384–400. https://doi.org/10.1177/1461444815605463

CCI. (2024). *The fraud of plastic recycling*. Centre for Climate Integrity. https://climateintegrity.org/plastics-fraud

Cowger, W., Willis, K. A., Bullock, S., Conlon, K., Emmanuel, J., Erdle, L. M., Eriksen, M., Farrelly, T. A., Hardesty, B. D., Kerge, K., Li, N., Li, Y., Liebman, A., Tangri, N., Thiel, M., Villarrubia-Gómez, P., Walker, T. R., & Wang, M. (2024). Global producer responsibility for plastic pollution. *Science Advances*, *10*(17), eadj8275. https://doi.org/10.1126/sciadv.adj8275

Dancing Fox. (2021). *The Break Free from plastic communications cookbook Asia edition 2021*. https://drive.google.com/file/d/1M7o_qabrs1HowecgdW5LOg4Ddu5PsXsn/view?usp=sharing&usp=embed_facebook

de Bakker, F. G. A. (2015). Online activism, CSR and institutional change. In J. Uldam & A. Vestergaard (eds), *Civic media and social engagement* (pp. 23–43). Palgrave Macmillan.

den Hond, F., & de Bakker, F. G. A. (2007). Ideologically motivated activism. *Academy of Management Review*, *32*(3), 901–924.

Flyverbom, M., Leonardi, P., Stohl, C., & Stohl, M. (2016). The management of visibilities in the digital age. *International Journal of Communication*, *10*(0), 98–109. https://ijoc.org/index.php/ijoc/article/view/4841

Friedman, M. (1999). *Consumer boycotts*. Routledge.

Gibbs, M., Meese, J., Arnold, M., Nansen, B., & Carter, M. (2015). #Funeral and Instagram: Death, social media, and platform vernacular. *Information, Communication & Society*, *18*(3), 255–268. https://doi.org/10.1080/1369118X.2014.987152

Hess, D. J. (2007). *Alternative pathways in science and industry: Activism, innovation, and the environment in an era of globalization*. MIT Press.

Joshi, S. (2020). Coca-Cola refuses to give up plastic bottles saying customers still want them. *VICE*. https://www.vice.com/en/article/coca-cola-refuses-to-give-up-plastic-bottles-saying-customers-still-want-them/

Kavada, A., Askanius, T., Kaun, A., Mattoni, A., & Uldam, J. (2023). Tying up Goliath: Activist strategies for confronting and harnessing digital power. In *State of power 2023-digital power* (pp. 96–115). Transnational Institute. https://www.tni.org/files/2023-04/Tying%20up%20Goliath.pdf

Langer, A. I., & Gruber, J. B. (2021). Political agenda setting in the hybrid media system: Why legacy media still matter a great deal. *The International Journal of Press/Politics*, *26*(2), 313–340. https://doi.org/10.1177/1940161220925023

Lekakis, E. J. (2017). Culture jamming and brandalism for the environment. *Popular Communication*, *15*(4), 311–327. https://doi.org/10.1080/15405702.2017.1313978

Lekakis, E. J. (2022). *Consumer activism: Promotional culture and resistance*. SAGE.

Lievrouw, L. A. (2023). *Alternative and activist new media*. John Wiley & Sons.

Lounsbury, M., Ventresca, M., & Hirsch, P. M. (2003). Social movements, field frames and industry emergence: A cultural-political perspective on US recycling. *Socio-Economic Review*, *1*(1), 71–104. https://doi.org/10.1093/soceco/1.1.71

Lyon, T. P., & Montgomery, A. W. (2013). Tweetjacked: The impact of social media on corporate greenwash. *Journal of Business Ethics*, *118*(4), 747–757. https://doi.org/10.1007/s10551-013-1958-x

MacBride, S. (2013). *Recycling reconsidered: The present failure and future promise of environmental action in the United States*. MIT Press.

Mah, A. (2021). Future-proofing capitalism: The paradox of the circular economy for plastics. *Global Environmental Politics*, *21*(2), 121–142. https://doi.org/10.1162/glep_a_00594

Micheletti, M., Stolle, D., & Follesdal, A. (2011). *Politics, products, and markets: Exploring political consumerism*. Transaction Publishers.

Montgomery, A. W., Lyon, T. P., & Barg, J. (2023). No end in sight? A greenwash review and research agenda. *Organization & Environment*, *37*(2), 221–256. https://doi.org/10.1177/10860266231168905

Mulholland, L., & Stern, M. (2021). Panorama: Coca-Cola's 100 billion bottle problem. [Broadcast]. *BBC One*. https://www.bbc.co.uk/programmes/m0010zxs

Powell, W. W., & DiMaggio, P. J. (2012). *The new institutionalism in organizational analysis*. University of Chicago Press.

Pulver, S. (2007). Making sense of corporate environmentalism: An environmental contestation approach to analyzing the causes and consequences of the climate change policy split in the oil industry. *Organization & Environment*, *20*(1), 44–83. https://doi.org/10.1177/1086026607300246

Rosch, C. (2024). The bloody turtle video that sparked a plastic straw revolution. *BBC Future*. https://www.bbc.com/future/article/20240402-the-turtle-video-that-sparked-a-plastic-straw-revolution

Thomas, D. (2020, January 21). Davos 2020: People still want plastic bottles, says Coca-Cola. *BBC News*. https://www.bbc.com/news/business-51197463

Uldam, J. (2016). Corporate management of visibility and the fantasy of the post-political: Social media and surveillance. *New Media & Society*, *18*(2), 201–219. https://doi.org/10.1177/1461444814541526

Van Dijck, J. (2013). *The culture of connectivity: A critical history of social media*. Oxford University Press.

van Laer, J., & van Aelst, P. (2010). Internet and social movement action repertoires. *Information, Communication & Society*, *13*(8), 1146–1171. https://doi.org/10.1080/13691181003628307

Yin, R. K. (2017). *Case study research and applications: Design and methods*. SAGE Publications.

Chapter 10

Greening the Platforms or Organisational Bullshit? A Typology of Empty and Misleading Environmental Communication by Global Platform Companies

Meri Frig[a], *Olga Dovbysh*[b] *and Salla-Maaria Laaksonen*[b]

[a]Aalto University and University of Helsinki, Finland
[b]University of Helsinki, Finland

Abstract

In this chapter, we identify and analyse how 'organisational bullshit', defined as empty, misleading or meaningless talk that is disconnected from organisational practices and realities, manifests in the corporate environmental communications of global platform companies. Our data include corporate sustainability reports from platform companies based in the United States, Europe, China and Russia. Building on theories of bullshit discourse in organisations, we reveal four types of organisational bullshit found in these reports: abstraction, distraction, reputation borrowing and stating the obvious, which were surprisingly similar across the four regions covered. These types stress how platform companies discursively obscure the climate impact of platform companies while promoting and exaggerating specific environmental activities and their significance. Some of the bullshit types are especially fitting for digital platform companies; as mediators of environmental communication and activism, they have unique opportunities for distraction and reputation borrowing. We argue that organisational bullshitting surrounding environmental sustainability distracts stakeholders from the actual problems and remedies but also propagates narratives of greener futures achieved with the help of the platform companies.

Platforms and the Planet, 167–182

doi:10.1108/978-1-83662-172-020261011

Keywords: Corporate social responsibility; sustainability reporting; global platform companies; organisational bullshit; critical CSR; environmental responsibility

Introduction

While the societal roles and responsibilities of platform companies are increasingly being scrutinised in public debate (e.g., Hill, 2020), their environmental responsibilities are receiving much less attention despite increasing knowledge of the mounting environmental impact of digitalisation (Freitag et al., 2021; Kuntsman & Rattle, 2019).

This impact is reported in corporate sustainability reports. Platform companies have adopted sustainability reporting practices similar to those of other industries. As data, corporate sustainability reports are a form of corporate social responsibility (CSR) communication through which companies report their activities, at the same time, construct, negotiate and modify what is meant by a socially responsible organisation (Christensen & Cheney, 2011). These reports entail future-orientated CSR communication, which, it has been argued, is performative and directs action towards aspired goals (Christensen et al., 2013, 2021; Schoeneborn et al., 2020). However, it can also be critically examined as a form of ideological persuasion (Mason & Mason, 2012) and as a way to postpone dealing with complex issues to an unknown time in the future (Tuan et al., 2024). More broadly, reporting is utilised to garner societal acceptance of platform companies and discursively construct these companies' desired realities and societal roles (Berger & Luckmann, 1966; Suchmann, 2005).

In this chapter, we build on theories of bullshit discourses in organisations (Christensen et al., 2019; Spicer, 2013, 2017, 2020) to increase our understanding of the reproduction and institutionalisation of organisational bullshit in corporate environmental communication. The term 'bullshit' that we use in analysing corporate discourses is more specific than the colloquial understanding of the term; that is, it does not refer to untrue, foolish or rude talk. Instead, it has been examined in the organisational communication literature as empty, misleading or meaningless talk disconnected from actual organisational practices and realities (Christensen et al., 2019; Spicer, 2013, 2017, 2020). Thus, bullshitting is communication characterised by unclarity, vagueness and indifference to truth (Cohen, 2012). We argue that it is important to recognise forms of environmental bullshit not only in the study of organisations and the environment but also for organisations themselves. For stakeholders and society, recognising bullshit helps in recognising when organisations are not telling the truth and are perhaps trying to avoid accountability. However, organisations can also benefit from recognising empty misleading jargon that easily develops into an industry-wide practice with little actual meaning-making ability. For example, being sensitive to bullshit could prevent significant wrongdoings and failures (McCarthy et al., 2020). Further, we argue that it is important for organisations to understand forms of organisational bullshit to create and engage in convincing and meaningful communication about how they interact with planetary boundaries.

Our empirical material consists of corporate sustainability reports by seven global platform companies – Meta (US), X/Twitter (US), Google (US), VK (Russia), Tencent (China), TikTok (China) and Spotify (Sweden) – spanning the years 2019–2023 – that is, after the publication of the Intergovernmental Panel on Climate Change's influential 'Special report on global warming of 1.5°C' (2018). Evidence from non-Western and authoritarian countries enables us to build a global view that expands the typical focus on so-called big tech companies based in the United States (US). It is important to explore understandings of environmental responsibility beyond the Western world. China, whose rapid economic development is coupled with heightened social and environmental concerns, and Russia, which hosts immense natural resources and burns more gas than any other country in the world (Bobylev & Perelet, 2013), are important examples. In this study, including authoritarian countries' perspectives on the global climate crisis allows us to reveal global platform companies' environmental discourses across different political and cultural contexts.

In what follows, we begin by turning to the theorisation of organisational bullshit before moving on to our empirical data and analysis. We identify four central types of bullshit in corporate sustainability reports of platform companies: abstraction, distraction, reputation borrowing and stating the obvious. We argue that organisational bullshitting surrounding environmental sustainability not only distracts stakeholders from the actual and serious problems as well as necessary actions but, in the case of platform companies, also builds narratives of futures in which these companies' digital services are an important part of a greener and more sustainable life.

Organisational Bullshit

'Bullshit' is described as obscure, empty or pretentious talk that is disconnected from organisational actions and realities (Spicer, 2013, 2020); in particular, it is prompted as an affective rejection of such talk (Christensen et al., 2019). However, unlike greenwashing, bullshitting is a social practice that can be 'accepted or even encouraged' (Spicer, 2020, p. 7) within a community or a specific industry/field and is therefore not necessarily specific to any one actor. In his theorisation of bullshitting, Spicer (2020) argued that whereas greenwashing is disapproved of, bullshitting is not necessarily discouraged but rather is popular or even expected in certain social settings – even if those engaging in it might not consciously recognise it. Based on past scholarship, organisational bullshit can be seen as a ubiquitous business practice (Christensen et al., 2019; Spicer, 2013). Because of its pervasiveness, it has become a natural and accepted part of certain practices, such as public relations, which stakeholders have learnt to read with a healthy amount of scepticism. CSR communication can be interpreted as including characteristics of organisational bullshit, specifically because much of it is closely connected to public relations and marketing (Christensen et al., 2019).

Christensen et al. (2019) distinguished between intentions to bullshit and the practice of bullshit: 'It is possible to emit bullshit without intending to do so' (p. 1591). Unlike lying or providing misinformation, such as greenwashing, organisational bullshit can be difficult to identify in, for example, sustainability reports. Corporate reports share not only information but also stories about organisational purposes and values, sustainability credentials and achievements, which are conveyed with the strategic aim of building a positive image and convincing a wide variety of stakeholders who may have very different expectations (Reinecke et al., 2017). Typically, sustainability reports contain future-orientated goals and aspirations that no stakeholder can object to. Accordingly, communication about companies' CSR efforts has been characterised as aspirational talk, or 'communication, which announces ideals and intentions rather than reflect actual behaviours' (Christensen et al., 2013, p. 2). Organisational climate targets can be examined as such future-orientated talk, which, perhaps ideally, can lead to substantive actions when internal or external audiences are inspired or provoked to play along with the ambitions (Christensen et al., 2013, 2021; Schoeneborn et al., 2020). These future-orientated action frames can coordinate actors and actions, but they can also be examined critically. For instance, symbolic or too-distant goals are often regarded as unambitious or even hypocritical, and organisations can be blamed for postponing controversies and difficult issues to the distant future (Christensen et al., 2020; Tuan et al., 2024).

Therefore, we argue that organisational bullshit in corporate sustainability reports obscures reliable information. Cohen (2006) pointed out obscure language that requires clarification, or 'unclarifiable unclarity,' as a characteristic of bullshit (p. 135). This is at the core of CSR communication, which is seen as a relational practice in which stakeholders seek to find out something meaningful about organisational practices by reading organisational statements. However, many stakeholders are sceptical about these statements, since information may be difficult to find, and the validity of such information can be difficult to prove. For instance, civil society groups often respond to corporate reports with counterarguments (Palazzo & Richter, 2005). Ironically, corporate sustainability reports are both official documents – a highly institutionalised and increasingly regulated form of communication – and unreliable, or at least inadequate, sources of information for many.

The aim of sustainability reporting, as well as standards, is to help stakeholders, from investors to consumers and civil society organisations, evaluate the sustainability of companies. However, it has been questioned whether sustainability reports can offer comprehensive and accurate perspectives on corporate actions because they are increasingly long and complicated (Fasan, 2013). Moreover, evaluating the accuracy of the information provided becomes more difficult when it moves from financial information to more narrative accounts of corporate activities (Garcia Torea et al., 2020). Accounting research has identified considerable selectivity bias in corporate sustainability reporting, thereby showing the relevance of this reporting as a public relations practice rather than as a tool for providing accurate accounts of environmental and social impacts (Cho et al., 2012). Yet, for many actors, including scholars, these reports are among the few means available to gain even a limited understanding of what companies are doing and to study what kinds of discourses are becoming institutionalised in sustainability reporting in each field.

What is particularly alarming about the institutionalisation of organisational bullshit is that the issues that CSR communication addresses are the most serious problems of our time. Indeed, Christensen et al. (2019) maintain that bullshit obscures information about these problems and therefore fails to signal potential danger (Vaughnan, 1996, cited in Christensen et al., 2019). Because of this, it is important that we learn to understand in more detail how bullshit materialises in corporate environmental communication.

Reading Corporate Sustainability Reports

Corporate annual reporting is a communication practice that aims to provide comprehensive disclosures of financial, operational and often sustainability-related information to stakeholders of corporations (Lord, 2002). New regulations (e.g., European Commission, n.d.), climate targets (e.g., science-based targets) and initiatives such as the Task Force for Climate-Related Financial Disclosures are reshaping corporate reporting practices to address climate-related risks. Financial incentives for climate-conscious practices are becoming increasingly prominent, driven by the rising consensus that failing to meet current climate targets will create greater costs in the future (Waddock et al., 2024).

In the European Union, legislation requires large and listed companies to publish regular reports on how their activities impact people and the environment as well as to disclose the social and environmental risks they face. The new Corporate Sustainability Reporting Directive requires companies to report in more detail on their negative impacts on human rights and the environment (European Commission, n.d.). In the United States, no similar requirement exists; nevertheless, a clear majority of big corporations report on their CSR activities (Governance & Accountability Institute, 2023). In Russia, there are no strict requirements for CSR reporting. However, the number of companies publishing these reports has been growing since 2001 (Zavrazhneva, 2024). In China, there are plans to increase sustainability-related environmental, social and governance (ESG) reporting to reach a disclosure rate of 70% by 2027 (Interesse, 2024).

Besides the seven aforementioned global platform companies that are the empirical focus of this chapter (Meta, X/Twitter, Google, VKontakte, Tencent, TikTok and Spotify), we checked the websites of the parent companies of Google (Alphabet) and TikTok (ByteDance) but found that neither published sustainability reports. Our choice of companies aims to provide an account of the CSR discourses used by major platform companies, such as Meta and Google, as well as to explore the corresponding discourses in non-US contexts and companies, represented by major Chinese platform companies (Tencent, TikTok), the largest Russian social media company VKontakte and the European music streaming company Spotify. Our empirical material consists of these companies' corporate sustainability/ESG reports published between 2019 and 2023. However, for the Chinese platform TikTok, our analysis is limited to the available materials: a small 'sustainability' subsection on its website and a single blog post announcing ByteDance's carbon neutrality commitment.

We began our analysis by mapping the problem statements, solutions, achievements and commitments mentioned in the platform reports. While doing this, we paid attention to the repetitive and vague language used across the reports. This observation led us to adopt a more critical perspective for our subsequent round of analysis, and we then grouped our observations from the first round to identify the types of organisational bullshit present in the data. Following Christensen et al. (2019), our focus was on the organisational bullshit itself, not on the intentions or credentials of the communicators. Based on our analysis of the reports' content, we identified four types of organisational bullshit: *abstraction, distraction*, *reputation borrowing* and *stating the obvious*.

Abstraction

The language used in corporate sustainability reports is characteristically vague and so arguably avoids making statements on difficult, complex, controversial or practice-orientated actions or explicit or specific commitments. Abstracted statements and pseudo-commitments (see also Ahmed, 2012) are expressed through empty language, metaphors and future-oriented language.

To begin with, reports are filled with statements that genuinely do not say anything, such as the following:

> We embrace our responsibility as a global company to address the climate challenge that impacts us all. (Meta, 2021, p. 7)
>
> We believe that every business has the opportunity and obligation to protect our planet. (Google, 2022, p. 2)
>
> A meaningful sustainability approach is good for business, our communities, and the world around us. (ByteDance, 2023)

Such statements are dressed up as commitments, even though their actual purpose seems to be to acknowledge the importance of environmental sustainability on a general level. Thus, they are not commitments to concrete action and do not offer much to hold the companies accountable for. It is difficult to evaluate, for example, the fulfilment of statements such as 'We are committed to protecting what is truly important: The well-being of people and our planet' (Meta, 2023, p. 8) and 'We're committed to protecting the environment, reducing our carbon footprint, and using our service to make a positive difference in the world', as Twitter (before being rebranded as X) announced in its 2020 Global Impact Report (p. 6).

Another form of abstracted expression in corporate reports involves future-orientated language – that is, corporate aspirations, goals and forward-looking statements that do not necessarily obligate companies to deliver them: 'We are committed to reducing our climate impact and raising awareness to promote positive change' (Spotify, 2023, p. 6). Some platform companies explicitly state such non-commitments at the end of their reports, as in the following example:

> Statements regarding targets, goals and commitments are aspirational and may also be based on estimates and assumptions under developing standards that may change in the future. As such, no guarantees or promises are made that they will be met or successfully executed, and actual results may differ, possibly materially. (Meta, 2023, p. 44)

In a similar manner, the sustainability reports of the platform companies frequently use metaphors, such as being 'just at the beginning of the journey' (Meta, 2021, p. 7; Spotify, 2023, p. 9). Here, the journey metaphor signals that even if a company is not acting in a sustainable way at the moment, stakeholders should expect it to become more sustainable in the future. At the same time, the companies state that accomplishing their future goals is not easy, as in the following example:

> We know this will be a challenging journey and we expect our progress towards this goal to be nonlinear. While our total GHG emissions increased from 2020 to 2021, we're committed to achieving GHG reductions throughout our operations and value chain. (Google, 2022, p. 3)

The vague commitments, metaphors and future-orientated language commonly found in the sustainability reports can be interpreted as 'aspirational talk' that reflects positive intentions rather than a thorough, critical analysis of organisational impacts. In general, vague commitments have been found to be common in organisations addressing social problems, creating an illusion that organisations can be imagined as subjects with commitments to such problems (Ahmed, 2012). However, critical reading reveals that such evasiveness can also be viewed as empty talk that is characteristic of organisational bullshit. This refers to obscure statements whose truth value cannot be negated or proved; these statements build a discursive connection between the platform company and the overall idea of environmental action but remain on a very high and abstract level.

Distraction

When the practical aim of corporate sustainability reports is to convince shareholders and other stakeholders of their good practices and intentions, companies seek to bring forth all the actions – the big and the symbolic – that make them seem sustainable. A close reading of the reports shows an emphasis on emissions reduction, as dictated by reporting standards, as well as on more symbolic actions that are meant to demonstrate care about the state of the planet. It can be argued that these smaller, more symbolic actions distract the reader from the big picture – that is, the relative weight of each action and what kinds of actions are meaningful and required at this time.

For example, Twitter 'stopped offering single-use plastic beverage bottles' and 'introduced 100% recycled content mailer packages in our San Francisco mailroom' (Twitter, 2020, p. 5). Meta installed beehives at its three offices, built up a green roof and a park in its headquarters and launched an app to track office food and waste reductions (Meta, 2021, 2023). Spotify says that it 'listen[s] to our planet's needs and take[s] the right actions' by, among other things, introducing meat-free days at its offices (Spotify, 2019, p. 26), while VK refuses to launch balloons made of non-recyclable materials or materials with long degradation times (VK, 2022). Although many of these actions are valuable, an emphasis on smaller, symbolic sustainability actions distracts the reader from questions such as what the major environmental responsibilities are of platform companies and where they are in addressing this.

We argue that these kinds of statements can be viewed as types of organisational bullshit, since they take focus away from the role and impact of the platform business itself. Although reducing meat consumption is essential in general, we lack better knowledge of how platform companies' business models contribute to the climate crisis. Distraction occurs in the descriptions of the various ways in which the concern of these companies for the environment is shown, such as in small actions in daily business operations, but arguably also in events that 'celebrate sustainability' and promote the platform brands as environmental ambassadors. For example, Spotify (2022) 'hosted climate sessions at local events, such as the Way Out West festival in Sweden' (p. 15), while Meta and Twitter 'celebrate' Earth Day and Earth Week. Google offices celebrate Earth Day and World Environment Day by hosting annual events at which they share 'sustainable living tips' on 'how Googlers can be sustainable on campus' (Google, 2020, p. 51) and host events with sustainability experts on the Talks at Google YouTube channel.

Hence, what is problematic about distraction as a form of organisational bullshit is that the biggest sources of emissions and resource usage do not receive the attention that would be called for. The platform companies also calculate and disclose their emissions, but these calculations typically receive less space than distracting positive discourses about the companies' incremental climate actions and beautiful photographs of nature (see also Kassinis & Panayiotou, 2018).

Reputation Borrowing

The reports frequently seek to associate the organisations with reputable and virtuous actors and organisations. By doing this, they *borrow the reputations of others*, thereby 'gaining respect by association' (Dowling, 2004, p. 23). Reports include descriptions of and links to reputable climate advocates, which can be institutions, organisations or smaller actors, such as podcasters and content producers.

Of course, companies involved with philanthropy and cause-related marketing are known to seek collaborations with non-profits, celebrities and so forth to improve their reputations and brand image (e.g., Erdogan, 1999). Similarly, corporate sustainability reports include descriptions of actors as well as actions through which the organisations have collaborated in some ways. Partnerships can include practical collaborations that promote forms of climate action or that do not require much involvement, such as 'providing ad credit' for a research study (Meta, 2021, p. 77).

Platform companies are well-positioned to take advantage of reputation borrowing, since simply by providing a platform for environmental advocates, they can argue that they have enabled important work. These companies emphasise this role in their reports; for example, Meta (2021) states that it 'create [s] solutions that connect people not only with each other but also with resources, information, tools and opportunities' (p. 65) and Tencent (2022) claims to 'focus on leveraging [its] digital technology and products to assist users, industries and society in transforming to a low carbon future' (p. 22).

Describing the platforms as sustainable due to their facilitating role in climate change communication can be considered problematic. The reports focus solely on specific examples of content that promotes environmental protection. For example, Spotify 'created a Climate Action hub on our platform where listeners could find a curated set of podcasts on climate change' (Spotify, 2021, p. 22). The same logic is highlighted by the 'Sustainability' section on TikTok's website under the broader section of 'TikTok for Good', in which 'the good our creators are doing all around the world' is presented as 'success stories.' However, as argued by the platform companies themselves (Gillespie, 2010; Napoli & Caplan, 2017), these companies' business models limit their role in curating the content. Therefore, these discourses disregard not only the climate impact of their technologies but also other negative consequences increasingly connected to social media platforms, ranging from information disorders to polarised discussion and addictiveness (e.g., Bhargava & Velasquez, 2021; Tucker et al., 2018).

Stating the Obvious

As corporate reports are used to convince stakeholders and to share stories about organisational values, such as protecting the environment, the reports provide quotes or statements that state the obvious, such as, 'Biodiversity is critical to ecosystem health' (Meta, 2021, p. 60) and 'The climate crisis is one of the most pressing issues of our time' (Spotify, 2021, p. 13). Many of these statements are often emphasised in bold letters in the reports, assuring readers that the corporations are aware of the state of the planet, thereby giving the impression that they also act accordingly.

Unlike not saying anything genuine, as discussed in the section on abstraction, stating the obvious conveys important facts. However, this can be interpreted as organisational bullshit because it can give the impression that awareness of or concern about the state of the planet is connected to organisational practices (Spicer, 2013, 2020). For instance, Spicer (2020) showed how some actors can 'hide their intentions of avoiding scrutiny by feigned fascination with the topic' (p. 12). Similarly, in the environmental communication literature, scholars have noted a proliferation of 'care discourses' and 'carewashing' – that is, the appropriation of the discourse of care by corporations to capitalise on the very crisis they have helped to create (The Care Collective, 2020, cited in Pezzullo, 2024, p. 2).

For example, Meta's 2021 sustainability report offers 'pushpin facts' throughout – that is, educational facts about the state of the planet and its people. In a way, this is in line with the platform company's stated ethos to elevate accurate information about climate change (e.g., Meta, 2023, p. 31). Stating the obvious is also a way to legitimise and provide background for the actions that a platform company has reportedly taken. Meta's sustainability report (2021) states that 'it is estimated that nearly two billion adults know little to nothing about climate change' (p. 66), emphasising the stated importance of its role in providing a platform for information and debates about the environment.

Finally, we argue that stating the obvious differs in one way from the other forms of organisational bullshit described above: there is intrinsic worth in doing this, such as indicating or clarifying organisational stances on the environment at a time characterised by polarisation and politicking over environmental action and corporate responsibilities. Namely, one of the existential challenges in environmental communication is that obvious facts remain repeatedly challenged. Returning to Christensen et al.'s (2019) argument that organisational bullshit obscures information about some of the most significant problems of our time, it must be noted that it remains crucially important to explicitly acknowledge the state of the planet. The challenge for environmental reporting in particular is to convincingly lean on this obvious fact to further justify organisational practices and drive organisational transformation.

Concluding Thoughts

In this chapter, we examined how global platform companies in the United States, Europe, China and Russia employ various corporate bullshit discourses while presenting their activities related to environmental sustainability in their official corporate sustainability reports. While the sustainability reports provide accounts of what the organisations have done and aspire to do to address the natural environment, they demonstrate a genre of environmental communication whose truthfulness and sincerity are frequently questioned (Ahmed, 2012; Lock & Seele, 2016). Employing a critical CSR approach and theories of organisational bullshit, we revealed four types of organisational bullshit – or alternatively, 'sustainability bullshit'– that stress how platform companies discursively reduce and blur the environmental impact of their activities while simultaneously exaggerating their own environmental activities and their significance.

Our study extends research on organisational bullshit (Christensen et al., 2019; Spicer, 2013, 2017, 2020) by identifying ways in which organisational bullshit is manifested in sustainability reporting – namely, providing *abstracted statements and commitments* that avoid difficult and complex topics by using vague language, *distracting* the reader from the environmental impact of the platform business itself, *borrowing the reputations* of reputable environmental advocates and *stating the obvious* to signify that care or awareness about the state of the planet is connected to organisational practices. We argue that understanding and recognising these types of organisational bullshit are highly important if we accept the claim that environmental communication is performative and, ideally, can lead to better practices and transformative change. Empty and obscure communication might be performative, but in less concrete and powerful ways when confronted with the institutional logics of neoliberal capitalism (see Fleming & Banerjee, 2016). This kind of environmental communication can be considered *non-performative*, as it does not necessarily commit the companies or their members to anything, nor does it guarantee their support for these commitments (Ahmed, 2012).

Since sustainability reporting has become a 'must-have' in CSR communication, companies produce voluminous and nicely written texts on their environmental commitments, values, aspirations and promotional activities, giving an impression of their environmental responsibility yet complicating the process of being held accountable. In Sara Ahmed's words, 'we need to refuse to read such commitments as performatives, as if they bring into effect that which they name' (2012, p. 121). However, bullshit in organisational communication has been characterised as pervasive yet difficult to pinpoint; it has been noted that it demonstrates an intuitive reaction to talk deemed untrustworthy (Christensen et al., 2019). It must be added that when analysing corporate environmental communication, this reaction can result in, for example, following the contrary goals of profit maximisation and environmental ambition. As can be expected, the platform companies announce in the sustainability reports that mitigating climate change and reducing emissions is a priority, but the first priority – business growth – remains drastically contradictory with the stated priority and is rarely highlighted. Against this backdrop, it is possible that, in the context of sustainability reports, even efficient and impactful environmental actions can be faced with scepticism.

In this study, we included companies from different parts of the world in our data, based on diverse political and economic regimes. While our goal was not to perform a systematic comparative analysis of the different companies, we noted differences in the reports in terms of the breadth and levels of detail in which actions were reported. Yet, the discourses and bullshit types they mobilised were surprisingly similar. Although the reasons for this are beyond the scope of this study, the findings demonstrate that providing aspirations, goals and values in corporate communication is a primary way for corporations around the world to address their environmental responsibility. Corporate reports are meant to serve as 'reference documents' that can be used to achieve such goals – commitments, for example, do not mean anything in themselves but must be accompanied by other actions (Ahmed, 2012, p. 120). 'Aspirational talk' in reports is believed to enable the exploration of those other actions (Christensen et al., 2013), especially when managers are unsure of what those other future actions can be (Spicer, 2020). This uncertainty has been argued to encourage the use and prominence of organisational bullshit (Spicer, 2020).

However, it is precisely the ubiquitousness of the identified discourses that makes it warranted to call them corporate bullshit. According to André Spicer (2020, p. 1; p. 20), 'playing the bullshit game' causes it to expand and dominate organisations and industries – unless it backfires (i.e. when 'what appeared as legitimate organizational processes are shown to be illegitimate'). That is, if we examine the bullshit as non-performative statements that each company must have and produce, sustainability bullshit is currently a widely approved practice unlike greenwashing, which we have distinguished as a separate yet related concept. While similar discourses have been identified in connection to cases of corporate greenwashing (e.g., Kassinis & Panayiotou, 2018; Mason & Mason, 2012), some of the bullshit types are specific for the field of digital platform companies, since by being the mediators of environmental communication and activism, they have unique opportunities for distraction and reputation borrowing. Further research, preferably inside companies, is needed to achieve a broader understanding of the potential national nuances in platform companies' environmental strategies (see also Pantti & Karell, this volume). The reluctance of companies to allow access to their internal activities is a major obstacle.

Although we accept that the goals of corporate sustainability reporting serve as a strategic tool to maintain and build a positive image as well as to achieve a competitive advantage (Rasche et al., 2017), we nevertheless stress the need to problematise the institutionalisation of organisational bullshitting in corporate environmental communication. The bullshit types discussed above can not only distract organisational stakeholders but also pose challenges for organisations to address these issues with the seriousness they require (Spicer, 2017). It therefore follows that it is important to highlight the urgent need for transparent, open-access, externally verified data on platforms' environmental impacts, as well as for clear and achievable global and national policies addressing the environmental responsibilities of global platform companies.

Acknowledgements

This work has been supported by The Media Industry Research Foundation of Finland, Kone Foundation and the Research Council of Finland [grant number 332751].

References

Ahmed, S. (2012). *On being included: Racism and diversity in institutional life*. Duke University Press.

Berger, P. L., & Luckmann, T. (1966). *The social construction of reality*. Doubleday.

Bhargava, V. R., & Velasquez, M. (2021). Ethics of the attention economy: The problem of social media addiction. *Business Ethics Quarterly*, *31*(3), 321–359. https://doi.org/10.1017/beq.2020.32

Bobylev, S., & Perelet, R. (2013). *Sustainable development in Russia*. Russian–German Environmental Information Bureau.

ByteDance. (2023). *ByteDance commits to operational carbon neutrality by 2030*. https://www.bytedance.com/en/news/64101e0917013dda5be8acc6

Cho, C., Michelon, G., & Patten, D. (2012). Impression management in sustainability reports: An empirical investigation of the use of graphs. *Accounting and the Public Interest*, *12*(1), 16–37. https://doi.org/10.2308/apin-10249

Christensen, L. T., & Cheney, G. (2011). Interrogating the communicative dimensions of corporate social responsibility. In Ø. Ihlen, J. F. Bartlett, & S. May (Eds.), *The Handbook of communication and corporate social responsibility* (pp. 491–504). John Wiley & Sons.

Christensen, L. T., Kärreman, D., & Rasche, A. (2019). Bullshit and organization studies. *Organization Studies*, *40*(10), 1587–1600. https://doi.org/10.1177/0170840618820072

Christensen, L. T., Morsing, M., & Thyssen, O. (2013). CSR as aspirational talk. *Organization*, *20*(3), 372–393. https://doi.org/10.1177/1350508413478310

Christensen, L. T., Morsing, M., & Thyssen, O. (2020). Timely hypocrite? Hypocrite temporalities in CSR communication. *Journal of Business Research*, *114*, 327–335. https://doi.org/10.1016/j.jbusres.2019.07.020

Christensen, L. T., Morsing, M., & Thyssen, O. (2021). Talk–action dynamics: Modalities of aspirational talk. *Organization Studies*, *42*(3), 407–427. https://doi.org/10.1177/0170840619896267

Cohen, G. A. (2006). Deeper into bullshit. In G. L. Hardcastle & G. A. Reisch (Eds.), *Bullshit and philosophy* (pp. 117–135). Open Court.

Cohen, G. A. (2012). Complete bullshit. In M. Otuska (Ed.), *Finding oneself in the other* (pp. 94–114). Princeton University Press.

Dowling, G. R. (2004). Corporate reputations: Should you compete on yours? *California Management Review*, *46*, 19–36. https://doi.org/10.2307/41166219

Erdogan, B. Z. (1999). Celebrity endorsement: A literature review. *Journal of Marketing Management*, *15*(4), 291–314. https://doi.org/10.1362/026725799784870379

European Commission (n.d.). *Corporate sustainability reporting*. https://finance.ec.europa.eu/capital-markets-union-and-financial-markets/company-reporting-and-auditing/company-reporting/corporate-sustainability-reporting_en. Accessed on December 18, 2024.

Fasan, M. (2013). Annual reports, sustainability reports and integrated reports: Trends in corporate disclosure. In C. Busco, M. Frigo, A. Riccaboni, & P. Quattrone (Eds.), *Integrated reporting* (pp. 41–57). Springer. https://doi.org/10.1007/978-3-319-02168-3_3

Fleming, P., & Banerjee, S. B. (2016). When performativity fails: Implications for critical management studies. *Human Relations*, *69*(2), 257–276. https://doi.org/10.1177/0018726715599241

Freitag, C., Berners-Lee, M., Widdicks, K., Knowles, B., Blair, G. S., & Friday, A. (2021). The real climate and transformative impact of ICT: A critique of estimates, trends, and regulations. *Patterns*, *2*(9), 100340. https://doi.org/10.1016/J.PATTER.2021.100340

Garcia Torea, N., Fernandez-Feijoo, B., & De la Cuesta, M. (2020). CSR reporting communication: Defective reporting models or misapplication? *Corporate Social Responsibility and Environmental Management*, *27*(2), 952–968. https://doi.org/10.1002/csr.1858

Gillespie, T. (2010). The politics of 'platforms'. *New Media & Society*, *12*(3), 347–364. https://doi.org/10.1177/1461444809342738

Google. (2020). *Google environmental report 2020*. https://sustainability.google/reports/google-2020-environmental-report/

Google. (2022). *Google environmental report 2022*. https://sustainability.google/reports/google-2022-environmental-report/

Governance & Accountability Institute. (2023). *Sustainability reporting trends*. https://www.ga-institute.com/research/research/sustainability-reporting-trends/

Hill, S. (2020). Politics and corporate content: Situating corporate strategic communication between marketing and activism. *International Journal of Strategic Communication*, *14*(5), 317–329. https://doi.org/10.1080/1553118X.2020.1817029

Interesse, J. (2024, June 20). China releases ESG reporting standards for businesses. *China Briefing*. https://www.china-briefing.com/news/china-releases-esg-reporting-standards-for-businesses/

Kassinis, G., & Panayiotou, A. (2018). Visuality as greenwashing: The case of BP and Deepwater Horizon. *Organization & Environment*, *31*(1), 25–47. https://doi.org/10.1177/1086026616687014

Kuntsman, A., & Rattle, I. (2019). Towards a paradigmatic shift in sustainability studies: A systematic review of peer reviewed literature and future agenda setting to consider environmental (un)sustainability of digital communication. *Environmental Communication*, *13*(5), 567–581. https://doi.org/10.1080/17524032.2019.1596144

Lock, I., & Seele, P. (2016). The credibility of CSR (corporate social responsibility) reports in Europe. Evidence from a quantitative content analysis in 11 countries. *Journal of Cleaner Production*, *122*, 186–200. https://doi.org/10.1016/j.jclepro.2016.02.060

Lord, H. L. (2002). Annual reports: A literature review (1989–2001). *Journal of Technical Writing and Communication*, *32*(4), 367–389. https://doi.org/10.2190/28LM-3HQR-R5QM-FCAU

Mason, M., & Mason, R. D. (2012). Communicating a green corporate perspective: Ideological persuasion in the corporate environmental report. *Journal of Business and Technical Communication*, *26*(4), 479–506. https://doi.org/10.1177/1050651912448872

McCarthy, I. P., Hannah, D., Pitt, L. F., & McCarthy, J. M. (2020). Confronting indifference toward truth: Dealing with workplace bullshit. *Business Horizons*, *63*(3), 253–263.

Meta. (2021). *2021 sustainability report*. https://sustainability.fb.com/wp-content/uploads/2022/06/Meta-2021-Sustainability-Report.pdf

Meta. (2023). *2023 sustainability report.* https://sustainability.fb.com/wp-content/uploads/2022/06/Meta-2021-Sustainability-Report.pdf

Napoli, P. M., & Caplan, R. (2017). Why media companies insist they're not media companies, why they're wrong, and why it matters. *First Monday*, *22*(5). https://doi.org/10.5210/fm.v22i5.7051

Palazzo, G., & Richter, U. (2005). CSR business as usual? The case of the tobacco industry. *Journal of Business Ethics*, *61*, 387–401. https://doi.org/10.1007/s10551-005-7444-3

Pantti, M., & Karell, E. (this volume). *Climate justice in Big Tech's environmental reporting.*

Pezzullo, P. C. (2024). On environmental communication as a care discipline. *Environmental Communication*, *18*(1–2), 1–7. https://doi.org/10.1080/17524032.2023.2300361

Rasche, A., Morsing, M., & Moon, J. (2017). *Corporate social responsibility. Strategy, communication, governance.* Cambridge University Press.

Reinecke, J., van Bommel, K., & Spicer, A. (2017). When orders of worth clash: Negotiating legitimacy in situations of moral multiplexity. In C. Cloutier, J. Gond, & B. Leca (Eds.), *Justification, evaluation and critique in the study of organizations* (pp. 33–72). Emerald Publishing Limited. https://doi.org/10.1108/S0733-558X20170000052002

Schoeneborn, D., Morsing, M., & Crane, A. (2020). Formative perspectives on the relation between CSR communication and CSR practices: Pathways for walking, talking, and t(w)alking. *Business & Society*, *59*(1), 5–33. https://doi.org/10.1177/0007650319845091

Spicer, A. (2013). Shooting the shit: The role of bullshit in organisations. *Management*, *16*, 653–666. https://doi.org/10.3917/mana.165.0653

Spicer, A. (2017). *Business bullshit.* Routledge.

Spicer, A. (2020). Playing the bullshit game: How empty and misleading communication takes over organizations. *Organization Theory*, *1*(2). https://doi.org/10.1177/2631787720929704

Spotify. (2019). *Sustainability & social impact report 2019.* https://investors.spotify.com/governance/default.aspx

Spotify. (2023). *Equity & social impact report 2023.* https://investors.spotify.com/governance/default.aspx

Suchman, M. C. (1995). Managing legitimacy: Strategic and institutional approaches. *Academy of Management Review*, *20*(3), 571–610. https://doi.org/10.2307/258788

Tuan, A., Corciolani, M., & Giuliani, E. (2024). Being reassuring about the past while promising a better future: How companies frame temporal focus in social responsibility reporting. *Business & Society*, *63*(3), 626–667. https://doi.org/10.1177/00076503231182627

Tucker, J. A., Guess, A., Barberá, P., Vaccari, C., Siegel, A., Sanovich, S., Stukal, D., & Nyhan, B. (2018). Social media, political polarization, and political disinformation: A review of the scientific literature. In *Hewlett foundation report.* https://doi.org/10.2139/ssrn.3144139

Twitter. (2020). *2020 global impact report.* https://about.twitter.com/content/dam/about-twitter/en/company/global-impact-2020.pdf

VK. (2022). *Key ESG projects and developments.* Q3 2022. https://corp.vkcdn.ru/media/files/q322-esg-highlights.pdf

Waddock, S., Henriques, I., Linnenluecke, M., Poggioli, N., & Böhm, S. (2024). The paradigm shift: Business associations shaping the discourse on system change. *Business and Society Review*, *129*(2), 155–167. https://doi.org/10.1111/basr.12359

Zavrazhneva, A. (2024, August 14). How non-financial reporting is developing in Russia today. *RBC trends*. https://trends.rbc.ru/trends/green/66bc74ea9a7947e3f4d2445?from=copy

Chapter 11

Climate Justice in Big Tech's Environmental Reporting

Mervi Pantti and Elis Karell

University of Helsinki, Finland

Abstract

Climate change disproportionately affects already disadvantaged countries in the Global South, evoking questions about the fair and just distribution of responsibilities between different actors for tackling climate change. The first international convention on climate crisis, the United Nations Framework Convention on Climate Change, emphasised 'common but differentiated responsibilities and respective capabilities', implying larger responsibility of Global North actors, state actors and corporations for taking action. This chapter addresses five major Big Tech companies – Alphabet/Google, Amazon, Meta, Apple and Microsoft – as moral agents who, because of their capacity, can be held responsible for combating climate change. From the perspective of climate justice, the authors asked how Big Tech companies address their responsibility for mitigating the effects of climate change in their sustainability reports and how their discourses incorporate justice concerns. This study shows that despite the adoption of climate justice terminology, platform companies pay limited attention to climate justice issues, and their corporate activities in the Global South reflect their economic and technological interests. The developing countries in the Global South appear merely as a scene where platform companies promote their social and environmental responsibility.

Keywords: Climate justice; Big Tech; corporate sustainability reporting; responsibility; recognition; distributive justice; Global South; climate change

Platforms and the Planet, 183–200

doi:10.1108/978-1-83662-172-020261012

Introduction

Climate change disproportionately affects already disadvantaged countries in the Global South. The most profound consequences of climate change are faced by those countries who are least responsible for contributing to the climate crisis. This has evoked questions about how to distribute responsibilities fairly between different actors for addressing climate change. The first international convention on the climate crisis – the United Nations Framework Convention on Climate Change (UNFCCC, 1992) – acknowledged that the responsibility to address climate change is collective but emphasised 'common but differentiated responsibilities and respective capabilities'. Both in terms of *responsibilities* (in sharing responsibility for past, present and future activities) and *capabilities* (ability to act on climate change), the convention imposed larger responsibility on industrialised countries (Shue, 2014).

This chapter addresses five major Big Tech companies – Alphabet/Google, Amazon, Apple, Meta and Microsoft – as *moral agents* who can be held responsible for addressing climate change and limiting climate damage. For assigning moral responsibilities to corporations, two frameworks are distinguished: (1) corporations' backward-looking responsibility to rectify their historic omissions (Shue, 2017) and (2) forward-looking responsibility, which is not based on past liability but on the ability to act (Young, 2011). Based on the literature in platform studies and climate justice, we argue that moral responsibility can be assigned to platform companies because of *their capability for counteracting the impacts of climate change*, in other words, their unique power as well as publicly stated commitment to corporate social responsibility and environmental sustainability. Addressing Big Tech companies as moral agents involves taking a normative approach that emphasises what corporations *should* do to mitigate climate change (Hormio, 2017). The greater the harm a corporation has caused or the more resources it possesses, the more responsibility it can be seen to bear.

The studied platform companies are multinational companies (MNC), leading companies based on market capitalisation, dominate their markets globally and are integral to political discourse on sustainable development, characterised by technological solutionism. We asked how Big Tech addresses and frames its responsibility for mitigating the effects of climate change in the Global South in its sustainability reports. This question was inspired by the discussion on *climate justice*, which draws attention to the unequal burdens of climate change (Schlosberg & Collins, 2014). Big Tech companies provide an intriguing case because of their 'platform power', the penetration of digital platforms in 'different economic sectors and spheres of life' (van Dijck et al., 2019, p. 1). They have accumulated significant social and political value in addition to economic worth, as other actors (businesses, institutions, governments and non-governmental actors) have become dependent on their infrastructural services (Poell et al., 2023; van Dijck et al., 2019). Thus, by expanding their power in offering infrastructural services, they reconfigure industries and shape social and cultural practices.

Platform companies' expansion of power in social and political life is seen, for instance, in how they have presented themselves as strongly invested in solving global problems from climate change to global poverty (e.g., Gates, 2021; Zuckerberg, 2017). In particular, Facebook's (Meta) and Google's long-standing mission has been to overcome the digital divide between the Global North and the Global South, and by doing so allegedly enhance development in the Global South through their services, products and expertise (Kwet, 2019; Nothias, 2020). Critical research has drawn attention to how Big Tech's 'social good' projects that aim to create positive social impacts serve to expand the companies' power and markets. Accordingly, the concepts of digital colonialism or data colonialism have referred both to the colonisation of digital infrastructure (Kwet, 2022; Mwema & Birhane, 2024; Oyedemi, 2020) and to social life by pushing data and technology-driven solutions for defining and solving social problems (Couldry & Mejias, 2019; Monsees et al., 2023).

Big Tech companies have set ambitious emissions reduction targets and present themselves as leaders on climate action (Russell, 2023) in their public communication (Pantti et al., 2025). However, they are controversial actors in global climate change governance and action. Big Tech contributes to environmental destruction and human rights violations through its global supply chains, increasingly energy-intensive data infrastructure, extraction of raw materials, exploitation of labour and electronic waste (Brevini, 2022). They typically oppose environmental regulations (Influencemap, 2021; Milman, 2021) and rely on potentially ineffective carbon market projects (Calma, 2024) and renewable energy certificate purchases (O'Brien, 2024) to deflate their officially reported emissions.

In the Global South, Big Tech is engaged in numerous carbon market projects, mainly forest and soil restoration projects. Apple, for instance, has invested in the conservation of mangrove forests to protect 'local communities' against the impacts of climate change. As explained by Lisa Jackson, Apple's vice president of Environment, Policy and Social Initiatives, 'The fight against climate change is a fight for the communities around the world whose lives and livelihoods are most threatened by the crisis, and that's where we've focused our work – from Colombia to Kenya to the Philippines' (Apple Newsroom, 2022). A similar emphasis on climate equity can be seen in Google, Meta and Microsoft's joint effort Symbiosis Coalition (2024) to offset their greenhouse emissions by 'carbon projects that meet the highest quality bar for planet and people, integrating the most recent science and data on the climate impact of restoration while equitably involving and compensating Indigenous Peoples and local communities' (Symbiosis Coalition, 2024).

This chapter contributes to the broader discussion on the platform-planet relationship by examining the ways in which climate justice is articulated in Big Tech's environmental reporting. To date, the literature on corporate responsibility has paid scant attention to corporate actions and discourses relating to the Global South. We proceed by first introducing climate justice as our theoretical perspective. This is followed by a discussion of the moral responsibility of private platform companies to tackle the climate crisis. After introducing our data and methods, we illustrate how the platform companies address climate justice concerns in their reporting of environmental sustainability and what implications this has for mitigating the impacts of climate change and redressing existing inequalities.

Climate Justice: Different Responsibilities for Climate Change

The effects of climate change disproportionately burden countries of the Global South and are reflected in racialised post-colonial communities, Indigenous Peoples and low-income households (Narain, 2019; Sultana, 2022). The Intergovernmental Panel on Climate Change (IPCC) (Intergovernmental Panel on Climate Change, 2023) estimates that nearly half the global population lives in areas highly vulnerable to climate change and, consequently, face bigger risks of food insecurity, water scarcity and mortality from extreme weather and climate events. The concept of *climate justice* highlights the unequal impacts of climate change, as well as inequalities regarding climate change mitigation. For example, literature points out climate justice issues related to carbon markets and renewable energy – solutions that are central to Big Tech's climate action. Carbon market projects can impose significant costs on Global South countries (Pande, 2024) and often provide little benefit to host communities (Mathur et al., 2014). Climate justice concerns related to renewable energy technologies include land use, exposure to pollution and reduced access to water resources (Levenda et al., 2021).

By focusing on human rights, climate justice introduces alternative perspectives alongside the dominant technology- and market-oriented discourse on climate change (Newell et al., 2021). However, within the climate justice movement, a diverse set of understandings has emerged. Major non-governmental organisations (NGOs), policymakers and corporate actors typically advocate for non-transformative approaches to climate justice that do not demand a systemic change; that is, a shift away from the capitalist economy requiring continuous growth. More radical discourses that have emerged within grassroots movements challenge current political economic arrangements and power relations and demand broader socio-economic transformation (Newell et al., 2021; Schlosberg & Collins, 2014; Shawoo & McDermott, 2020; Sultana, 2022).

Climate justice is often seen to encompass overlapping dimensions of *recognitional justice*, *distributive justice* and *procedural justice* (Kortetmäki, 2016; Newell et al., 2021). We focus on recognitional and distributive justice since procedural justice is most pertinent in the political sphere. It brings forth the unequal ability of participation in decision-making and highlights the importance of sensitivity to already-existing inequalities (Newell et al., 2021).

Recognitional climate justice concerns the recognition of injustices and their causes, as well as the recognition of cultural and social differences (Newell et al., 2021; Schlosberg, 2012). Recognitional justice can draw attention to climate-related inequalities on a global or local level or inequalities based on gender, race, class or generational divides (Newell et al., 2021; Shawoo &

McDermott, 2020). Justice theorists argue that recognition of the social relations that create injustice is a prerequisite for distributive and procedural justice (Fraser, 2007; Schlosberg, 2012; Young, 2006). A lack of recognition of the climate vulnerability of specific groups can lead to inefficient approaches to climate action, with detrimental effects on those groups (Schlosberg, 2012). In the corporate discourse on sustainability, companies can, for example, recognise certain groups that are disproportionately vulnerable to climate change, the specific inequalities causing the disproportionate vulnerability and the nature of the disproportionate climate-related harms.

Distributive climate justice refers to material and economic inequalities in sharing the burdens of climate change. Because climate change exacerbates global inequality, responses need to include distributive principles that allocate mitigation, adaptation and compensation burdens in a just way (Sardo, 2023). Discussions of distributive climate justice involve identifying which costs and benefits or social goods and ills are being distributed, between whom and how (Newell et al., 2021). Typical corporate sustainability activities that relate to distributive climate justice include efforts to reduce the negative impacts of corporate operations on vulnerable communities, participating in climate change mitigation and adaptation efforts, and considering the climate justice implications of such efforts (e.g., renewable energy or carbon market projects).

Climate justice also draws attention to actors' different responsibilities and capabilities in addressing climate change (Newell et al., 2021). As we stated in the introduction, sharing responsibility can be based on an actor's *contribution* to the climate crisis, which means that the highest emitters should be assigned greater compensation burdens. On the other hand, distributing responsibility can also be based on an actor's *capability* to act on climate change regardless of their causal contribution (Shue, 1999, p. 537). We turn next to this subject by shifting the perspective to corporate moral responsibility in addressing justice concerns.

Moral Responsibility of Big Tech

The private sector is increasingly expected to act in the interests of society and the planet. In particular, large carbon corporations that are primary contributors to CO_2 emissions have been described as the primary agents who should compensate for climate change-related harm (Grasso, 2024; Grasso & Vladimirova, 2020; Shue, 2017; Umbers & Moss, 2020). The trend of demanding more from corporations is seen, for example, in the wave of climate litigation against large fossil fuel companies. Increasing stakeholder pressure has led major corporations to adopt more ambitious positions on climate issues, often framing themselves as leading actors in mitigating climate change (Russell, 2023; Schlichting, 2013).

Platform companies have more recently been seen as contributors to global climate change, rather than as part of the solution (Brenner & Hartl, 2021; Kuntsman & Rattle, 2019). We argue that the responsibility of platform companies for taking action to prevent and compensate for climate-related harm should not only be based on their contribution to the climate crisis but also on their power and capability to act (Hedlund, 2022; Young, 2006). We elaborate this through three principles of justice that have been proposed to distribute the burdens involved in addressing the climate crisis.

Two of the main *backward-looking principles* are the polluter pays principle (PPP) and the beneficiary pays principle (BPP) (e.g., Shue, 2014). The PPP contends that responsibility for addressing climate change should correlate with causal contributions to climate change. Based on the PPP, the concept of 'climate debt' has been raised within the climate justice movement, stressing that the historical responsibility lies with industrialised countries and their businesses (e.g., Singh & Bajwa, 2023). The BPP emerged from criticism of the PPP, stressing that responsibility should correlate with the benefits gained from carbon-intensive activities. Moreover, responsibility can be seen as being not only based on passively received benefits but also on active participation in a carbon-intensive global political and economic system (Sardo, 2023).

A third principle – the capacity principle or the ability to pay principle (APP) – assigns responsibility in proportion to an agent's capacity. It does not consider backward-looking issues but, instead, is a *forward-looking principle* by focusing on an actor's capability to act responsibly (van de Poel, 2015). The allocation of forward-looking responsibilities can be grounded in actors' economic resources, social connections and political power, including the capacity to influence public debate and political decision-making (Fraser, 2007; Hedlund, 2022; Kortetmäki & Huttunen, 2023; Young, 2006).

Thus, the moral responsibility of corporations can be distinguished through their resources and knowledge capabilities (APP principle) and benefits related to their areas of business (BPP principle). For example, some businesses benefit from the green transition (Dolšak & Prakash, 2022; Hulme, 2023). Consequently, Big Tech's moral responsibility can be found in the vast economic opportunities presented by climate change to develop new products and services that satisfy green markets. Google, Amazon, Meta, Apple and Microsoft provide a considerable variety of services built on transnational infrastructure (such as cloud computing systems) that sustain their dominance of the platform ecosystem. Platform power means that digital platforms can control and enhance sustainable practices in their supply chain. Moreover, as providers of information infrastructure, they shape public perceptions by facilitating information production and circulation about climate change.

Climate Justice in Corporate Sustainability Reports

Our aim was to determine how moral responsibility and global climate justice concerns are communicated in platform companies' sustainability reports. Does their framing of climate change incorporate inequalities in vulnerability to climate change and their own responsibility for mitigating these inequalities? Sustainability reports document corporations' climate action, and by making their policies and aspirations public, they allow a critical evaluation of their climate responsibility. While the content of corporate sustainability reports is to some extent dictated by reporting guidelines, such as the Global Reporting Initiative, companies still have significant freedom in deciding what information to include and how to present it (Pollach, 2018). We understand environmental reporting as a communicative action that is used to construct a positive image of a company, disseminate information and aspirations, communicate responsibility to stakeholders and the public, as well as draw the limits of that responsibility (e.g., Siltaoja & Onkila, 2013). Such talk is not without consequences, as it has implications for how we understand the responsibilities of businesses. It may shape business standards as well as give 'weapons' to stakeholders, including customers and advocacy groups, to critically examine the authenticity, effectiveness and long-term commitment of corporate claims and hold companies to their words (Christensen et al., 2013).

Apple and Google have published standalone sustainability reports for over a decade, whereas Amazon, Meta and Microsoft have started publishing them as late as 2020 or 2021. In order not to skew the material towards certain companies, we only included reports published since 2020. All sustainability reports were gathered from the companies' websites, resulting in a total of 24 reports (five reports each by Amazon, Apple, Google and Meta, and four reports by Microsoft). The body text of the reports was extracted and analysed in Atlas.ti.

First, we identified and coded mentions of Global South countries based on the definitions of developing countries by the UN Trade and Development (UNCTAD) (UN Trade and Development, 2024). Mentions that were not related to climate change or environmental issues (e.g., gender equality or labour issues) were excluded. The coding was grouped by continent, with the largest countries and high Human Development Index (HDI) countries separately, resulting in the codes Africa (every mention of Africa or African countries), Asia (excl. China, India, HK, TW, SG), China, Hong Kong/Taiwan/Singapore, India and Latin America and the Caribbean (LAC). The reports did not mention any Oceanian countries other than Australia and New Zealand.

Second, we identified and coded mentions of the term 'climate justice', as well as 'environmental justice', 'climate equity' and 'environmental equity'. We also reviewed every mention of the term 'climate change' for possible climate justice framings. The search resulted in 69 text passages directly addressing issues of climate justice.

The data were analysed using frame analysis. This choice was based on the premise that addressing an environmental issue depends on how the problem is defined. According to Entman (1993), a frame consists of a definition of a problem or issue, a causal attribution, a moral evaluation and a treatment. We focused on how the platform companies define the problem they are addressing, the perceived cause of this problem, what actions to remedy the problem are suggested or described and how the actions are justified.

Recognition of Injustice and Vulnerable Groups in Sustainability Reports

The texts that addressed climate justice were organised into the categories of *recognitional* and *distributive climate justice*. Recognitional justice manifests in understanding differences and acknowledging the diverse impacts of climate change on different societal and cultural groups. Recognition's central assertion is that social justice requires not only the just distribution of material resources but also an acknowledgement (Honneth, 1995). In a narrow sense, recognition is the process of making someone or something visible. The mediated public sphere, including corporate media analysed here, provides a forum for the politics of recognition.

The sustainability reports show that platform companies have adopted the language of climate justice in their sustainability discourse: the reports address climate justice in varying levels of detail, ranging from mentioning climate justice in passing to more elaborate accounts. Apple and Microsoft address climate justice at least once in every included report, while Amazon, Google and Meta refer to climate justice less often. References to climate justice are typically accompanied by defining certain groups as disproportionately vulnerable to climate change. The most commonly identified groups include Indigenous Peoples and people of colour (typically in the context of US ethnic minorities), as well as vaguely defined groups such as 'vulnerable communities', 'low-income and historically marginalised communities', 'economically distressed communities', 'underrepresented communities', 'overburdened communities', 'historically underserved communities', 'disadvantaged communities' and 'communities that are disproportionately affected'. Apple, for instance, when introducing their Impact Accelerator programme (which is part of Apple's Racial Equity and Justice Initiative), writes that 'we know we can help spur an economic recovery that tackles systemic barriers to opportunity and empowers local communities – especially the communities most vulnerable to the impacts of climate change' (Apple, 2021). As these examples show, the discourse on climate vulnerability and justice stays at a vague and abstract level, appearing as a discursive strategy to create an exaggerated impression of the companies' environmental responsibility using a carefully selected vocabulary (see Frig et al., this volume).

Notably, the reports hardly ever connect climate justice to the Global South. Fig. 11.1 shows the distribution of references to different countries in the reports. Unsurprisingly, the United States, India and China receive the most attention, as the companies' supply chains are located primarily in these countries. Outside of India and China, the attention given to Global South countries is sporadic and far less frequent. Out of African countries, only South Africa receives more than a few mentions, and in Latin America, Mexico and Brazil stand out as the most mentioned countries.

Instead, the reports address climate justice from the perspective of various geographically unspecific vulnerable groups and using the term 'local community', thus concealing global inequality as a cause of disproportionate climate change vulnerability. There are only a few exceptions: Microsoft refers to the 'least developed countries' (2023) and to the 'Global South' (2024) in the context

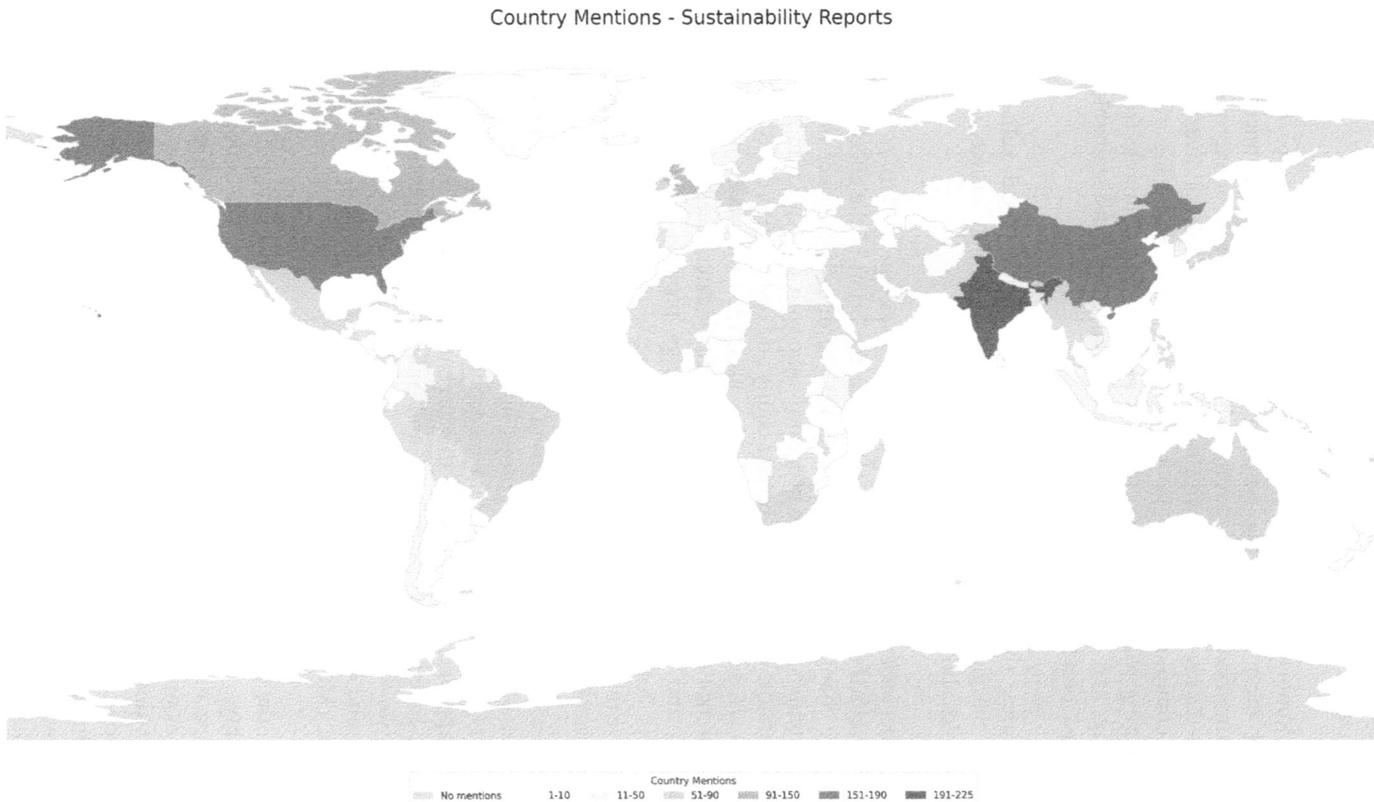

Fig. 11.1. References to Countries in Platform Companies' Sustainability Reports. *Source:* Produced by Erjon Skënderi.

of climate justice. In some texts, Global South countries are named when companies introduce their commitment to equity:

> In Colombia, it [making our work 'an even greater force for good'] meant partnering with local community, government, and conservation organizations, which help preserve mangrove forests deeply intertwined with the economic and cultural lives of their communities. And in Kenya, it meant bolstering the pioneering work of conservationists supporting sustainable tree farming and preserving savannas degraded by unsustainable land-use practices. (Apple, 2020)

In most cases, the reports primarily identify various local vulnerabilities (typically US marginalised groups) in the context of climate justice, as the following examples show:

> Consider *underrepresented communities: Low-income and historically marginalized communities* too often bear the brunt of the effects of climate change. So we're pursuing ways to directly consider these communities in our climate program – like with the Apple Impact Accelerator, which aims to bolster equity and foster opportunity for *Black-, Hispanic/Latinx-, and Indigenous*-owned businesses in the environmental sector. We're advocating for greater access to clean energy and investing in renewable energy projects in emerging markets. And we're targeting carbon removal projects that aim to improve the livelihoods of *local and Indigenous communities*. (Apple, 2022, our italics)

> We are committed to supporting climate solutions that take *impacted communities* and groups into consideration, advancing projects and partnerships that incorporate equity and justice in their approach to addressing this urgent issue. As we invest in nature-based carbon removal and infrastructure around the globe, we will also prioritize projects that create environmental, social, and economic benefits for *people most impacted by climate change*. (Meta, 2021, our italics)

Other than the injustices alluded to by the quotes above – that is, ethnic and racial inequality, poverty or historical marginalisation – the reports do not identify any specific injustices as causes of the groups' vulnerabilities to climate change. For example, climate injustices based on gender or age (women or younger generations being disproportionately affected by climate change) are not addressed. Moreover, the texts rarely detail *in what way* vulnerable groups are affected by climate change, except for a few mentions of disproportionate climate impacts relating to food, water access, health, pollution, standard of living, extreme weather and natural disasters.

In sum, climate justice largely remains at a level of acknowledgement in sustainability reports. It is brought forward as a problem, but typically without describing any specific remedies. Instead, the problem framing of climate injustice is linked to vague solutions, such as taking climate justice into account in corporate operations, without any specific distributive or other concrete actions stated, as the examples below show:

> We added several new elements to our standards too, including: [. . .] An environmental justice provision that encourages suppliers to assess and address environmental equity issues. (Amazon, 2023)
>
> We recognize that climate and environmental issues don't affect every community in the same way and that *we need to address environmental equity* as a broader issue. In fact, those most affected by climate change are those who have contributed the least to climate change. Renewable energy procurement can *help address these inequities*. (Microsoft, 2021)

Distributive Justice: Sharing the Burden and Taking Responsibility

The sustainability reports include a wide range of descriptions of corporate activities in the Global South that in some way concern the distribution of climate-related burdens. However, as we stated above, these activities are seldom framed as responding to climate injustice or inequality. There are only some exceptions in which corporate actions, including funding and investing, NGO partnerships or technological solutions, are framed as specific remedies to climate injustice, as shown in these examples:

> The Environmental Justice Data Fund, a collection of grants totaling nearly $9 million that aim to help frontline communities that have been historically underserved and disproportionately impacted by climate change and environmental injustice. (Google, 2023)
>
> Incorporating environmental justice principles into our carbon-free electricity procurement, we partnered with EKOenergy's Climate Fund to enable the installation of solar-powered refrigeration for a Kenyan fishing village, offering ice and clean water at 20% lower prices for 2,000 households and addressing food security and economic hardship by reducing fish harvest loss. (Microsoft, 2024)

In reports that mention Global South countries without explicitly addressing climate justice issues, two broad frames emerged: *corporate sustainability* and *global climate responsibility*. The first frame relates to the sustainability of companies' operations in the Global South. In *the corporate sustainability frame*,

CO_2 emissions and other environmental impacts (e.g., water use, waste, raw materials) caused by the company are defined as the problem. However, this problem definition only occurs indirectly through descriptions of already (or soon to be) taken corporate actions that are framed as solutions to the problem, as in Amazon's account:

> In India, 81% of last mile deliveries are made by delivery associates using personal two-wheel vehicles. To encourage DSP [delivery service partner] associates to convert their personal vehicles to electric, we offer exclusively sourced EV [electronic vehicle] deals from original equipment manufacturers as well as access to affordable financing. (Amazon, 2024)

The primary solutions in the corporate sustainability frame are reducing emissions through increased efficiency (e.g., decreasing energy use, developing more energy-efficient technology, recycling) and investing in renewable energy. Microsoft (2021), for example, states that '[i]n South Africa, we use atmospheric water generation to collect water out of the air to fill storage tanks for use in cooling our datacentres'. The corporate sustainability frame is predominantly associated with China, India and the rest of Asia – countries where a significant part of the companies' operations and supply chains are located.

In the more prominent frame of *global climate responsibility*, climate change and other environmental issues in the Global South are defined as the problem. This frame typically does not include any recognition of global inequalities that cause disproportionate climate and environmental vulnerabilities in Global South countries. In other words, the Global South does not play a significant role in the problem definition of this frame. Furthermore, the problem definitions lack any diagnoses of causes or evaluations of the problems. The most common remedies offered in the global climate responsibility frame, in order of prevalence, are partnerships (both NGO and businesses), increasing the availability of renewable energy, technological solutions, funding/investing in solutions (e.g., sustainability-related start-ups or initiatives) and conservation projects. While a wide variety of actions are framed as remedies, they can be divided into two main categories: distribution of capital (through partnerships, initiatives and investments) and making technological solutions more available (through development or investments). Both are illustrated in the following example of the global climate responsibility frame:

> We have also partnered with Conservation International in Kenya to protect and restore savanna with the Maasai Wilderness Conservation Trust and the Big Life Foundation in the iconic Chyulu Hills region. Our funds will help to repair damage caused by unsustainable land use practices that degrade grasslands and soils, endanger local wildlife, release carbon into the atmosphere, and deprive local communities of vital agricultural resources. (Apple, 2021)

While the global climate responsibility frame de facto concerns the distribution of climate-related burdens between major US companies and disadvantaged Global South recipients, this global inequality is rarely articulated as a justification for the actions explained in the reports. In most cases, no explicit justification is given. However, two main justifications appear somewhat frequently: wider social/economic benefit (of the described actions) and enabling other actors' sustainability. Social and economic benefits are virtually exclusive to the global climate responsibility frame, while actions in both frames are occasionally justified by enabling other actors' sustainability.

Conclusion

We set out to discuss why Big Tech should be assigned enlarged responsibilities for combating climate change and to acknowledge those who are most impacted by climate change. At the same time, we recognised the equally powerful potential of profit-driven global corporations to exacerbate the climate crisis and make the global majority increasingly vulnerable to its effects. The climate justice perspective adopted in this chapter is a normative idea that suggests corporations with the capacity to influence and act effectively should take responsibility. It is an argument commonly used to pressure private actors, primarily focused on their interests, to enhance their climate actions and to encourage governmental actors to enforce regulations on corporate sustainability.

Our analysis makes the critical observation that while platform companies have adopted the terminology of climate justice in their sustainability reporting, they pay little attention to the Global South. When environmental issues in the Global South are addressed, questions of power relations and climate policy (apart from renewable energy policy) are typically avoided. North–South asymmetries in climate responsibility and vulnerability are not addressed in the framing of carbon market projects and other local sustainability projects as treatments to climate change or company CO_2 emissions. Instead, the reports describe a wide range of sustainability activities in the Global South, and while almost all these activities are distributive, they are typically framed as responding to broadly defined problems, such as climate change, with no acknowledgement of the global inequalities. Consequently, the Global South appears merely as a scene in which platform companies promote their social and environmental responsibility to gain a positive public image, perpetuating colonial ideologies through othering.

However, the unequal effects of climate change are recognised as a problem, and the reports distinguish disproportionately vulnerable groups. These vulnerable groups are not defined based on inequalities between developed and developing countries. Instead, disproportionate climate vulnerability is primarily articulated as a feature of various geographically unspecific groups or ethnic (mostly US) minorities. Perhaps this is because the companies avoid engagement in highly contentious political issues, such as just climate funding for the Global South (Schneider, 2024), to shield themselves from criticism and cater to their audiences and customers in the Global North.

In line with the genre of platforms' corporate sustainability reporting (see Frig et al., this volume), problem framings that attribute responsibility to the company feature a consistent focus on remedies that are presented as already accomplished or in progress. The limited focus on problem definitions and causal attributions shows an 'avoidance of indictment', typical in the climate change communication of high-emitting large companies (Megura & Gunderson, 2022). While Big Tech's role in causing climate change is not nearly as significant as that of the fossil fuel industry, any extended problem definition and causal attribution in corporate communication would, to some extent, indict the economic system they benefit from. The corporate activities in Global South countries described in the reports reflect the economic and technological power of the companies, large investments in renewable energy, extensive partnerships and technological or data-based solutions. The heavy focus on technological solutions to address the climate crisis indicates the dominant techno-solutionist discourse identified in several studies (Iazzolino & Stremlau, 2024; Kuntsman & Rattle, 2019; Laaksonen & Frig, 2025; Megura & Gunderson, 2022).

As global efforts to combat the climate crisis continue to be insufficient, issues of climate justice, especially the Global North's responsibility to meet developing countries' climate needs, are becoming increasingly severe. We have argued that the socio-economic and political power of platform companies makes them uniquely positioned to have a significant contribution to global efforts to address climate change impacts and their unequal effects. While the private sector is recognised as an increasingly important actor in international climate policy, corporate discourses have not been examined in terms of the climate justice framings. Our research shows that, despite their strong public pro-climate stances and adoption of climate justice discourse, platform companies pay only limited attention to Global South countries, which reflects the interests of the global platform economy. We emphasise the need to study corporate discourse to understand how it contributes to the meaning-making of climate justice both globally and across different geographic regions. By defining their responsibilities, perceived problems and preferred solutions in their environmental reporting, Big Tech companies also reveal the limits of their responsibility, thereby opening avenues for holding them responsible for their contribution to climate change and its effects on human rights.

Acknowledgements

The authors would like to thank Erjon Skënderi for helping with the data. This work was supported by the Research Council of Finland [grant number 332751] and The Strategic Research Council (SRC) established within the Research Council of Finland [grant number 372226].

References

Amazon. (2023). *2022 Amazon sustainability report: Building a better future together*. https://sustainability.aboutamazon.com/

Amazon. (2024). *2023 Amazon sustainability report*. https://sustainability.aboutamazon.com/2023-amazon-sustainability-report

Apple. (2020). *2020 environmental progress report*. https://www.apple.com/environment/pdf/Apple_Environmental_Progress_Report

Apple. (2021). *2021 environmental progress report*. https://www.apple.com/environment/pdf/Apple_Environmental_Progress_Report

Apple Newsroom. (2022, April 21). *From farm to sea: Conserving mangroves to protect local livelihoods and the planet*. https://www.apple.com/newsroom/2022/04/conserving-mangroves-to-protect-local-livelihoods-and-the-planet/

Brenner, B., & Hartl, B. (2021). The perceived relationship between digitalization and ecological, economic, and social sustainability. *Journal of Cleaner Production*, *315*(5), 128128. https://doi.org/10.1016/j.jclepro.2021.128128

Brevini, B. (2022). *Is AI good for the planet?* Polity Press.

Calma, J. (2024, May 22). Big Tech thinks it can plant trees better than everyone else. *Verge*. https://www.theverge.com/2024/5/22/24162428/google-meta-microsoft-sales-force-symbiosis-coalition-nature-carbon-removal-forest-offsets

Christensen, L. T., Morsing, M., & Thyssen, O. (2013). CSR as aspirational talk. *Organization*, *20*(3), 372–393. https://doi.org/10.1177/1350508413478310

Couldry, N., & Mejias, U. A. (2019). Data colonialism: Rethinking big data's relation to the contemporary subject. *Television & New Media*, *20*(4), 336–349. https://doi.org/10.1177/1527476418796632

Dolšak, N., & Prakash, A. (2022). Three faces of climate justice. *Annual Review of Political Science*, *25*(1), 283–301. https://doi.org/10.1146/annurev-polisci-051120-125514

Entman, R. M. (1993). Framing: Toward clarification of a fractured paradigm. *Journal of Communication*, *43*(4), 51–58. https://doi.org/10.1111/j.1460-2466.1993.tb01304.x

Fraser, N. (2007). Re-framing justice in a globalizing world. In N. Fraser & P. Bourdieu (Eds.), *(Mis)recognition, social inequality and social justice* (pp. 29–47). Routledge.

Frig, M., Dovbysh, O., & Laaksonen, S.-M. (this volume). *Greening the platforms or organisational bullshit? A typology of empty and misleading environmental communication by global platform companies*.

Gates, B. (2021). *How to avoid a climate disaster: The solutions we have and the breakthroughs we need*. Vintage.

Google. (2023). *2023 environmental report*. https://www.gstatic.com/gumdrop/sustainability/google-2023-environmental-report.pdf

Grasso, M. (2024). Fossil fuel companies' duty of reparation: Why the industry must concur to foot the climate bill. *Globalizations*, *21*(4), 630–647. https://doi.org/10.1080/14747731.2023.2239566

Grasso, M., & Vladimirova, K. (2020). A moral analysis of carbon majors' role in climate change. *Environmental Values*, *29*(2), 175–195. https://doi.org/10.3197/096327119X15579936382626

Hedlund, M. (2022). Distribution of forward-looking responsibility in the EU process on AI regulation. *Frontiers in Human Dynamics*, *4*, 703510. https://doi.org/10.3389/fhumd.2022.703510

Honneth, A. (1995). *The struggle for recognition: The moral grammar of social conflicts*. The MIT Press.

Hormio, S. (2017). Can corporations have (moral) responsibility regarding climate change mitigation? *Ethics, Policy & Environment*, *20*(3), 314–332. https://doi.org/10.1080/21550085.2017.1374015

Hulme, C. (2023). *Corporate climate action, transnational politics, and world order*. Palgrave Macmillan.

Iazzolino, G., & Stremlau, N. (2024). AI for social good and the corporate capture of global development. *Information Technology for Development*, *30*(4), 626–643. https://doi.org/10.1080/02681102.2023.2299351

Influencemap. (2021). *Big Tech and climate policy*. https://influencemap.org/report/Big-Tech-and-Climate-Policy-afb476c56f217ea0ab351d79096df04a

Intergovernmental Panel on Climate Change. (2023). *Climate change 2023: Synthesis report*. https://doi.org/10.59327/IPCC/AR6-9789291691647.001

Kortetmäki, T. (2016). Reframing climate justice: A three-dimensional view on just climate negotiations. *Ethics, Policy & Environment*, *19*(3), 320–334. https://doi.org/10.1080/21550085.2016.1226238

Kortetmäki, T., & Huttunen, S. (2023). Responsibilities for just transition to low-carbon societies: A role-based framework. *Environmental Politics*, *32*(2), 249–270. https://doi.org/10.1080/09644016.2022.2064690

Kuntsman, A., & Rattle, I. (2019). Towards a paradigmatic shift in sustainability studies: A systematic review of peer reviewed literature and future agenda setting to consider environmental (un)sustainability of digital communication. *Environmental Communication*, *13*(5), 567–581. https://doi.org/10.1080/17524032.2019.1596144

Kwet, M. (2019). Digital colonialism: US empire and the new imperialism in the Global South. *Race & Class*, *60*(4), 3–26. https://doi.org/10.1177/0306396818823172

Kwet, M. (2022). *Digital colonialism and infrastructure-as-debt*. University of Bayreuth African Studies Online. https://papers.ssrn.com/sol3/papers.cfm?abstract_id=4004594

Laaksonen, S.-M., & Frig, M. (2025, in press). Narratives of indispensability and infrastructural solutionism by AI giants. In A. Mollen, S. Kannengießer, F. Jansen, & J. Velkova (Eds.), *AI infrastructures and sustainability*. Palgrave Macmillan.

Levenda, A. M., Behrsin, I., & Disano, F. (2021). Renewable energy for whom? A global systematic review of the environmental justice implications of renewable energy technologies. *Energy Research & Social Science*, *71*, 101837. https://doi.org/10.1016/j.erss.2020.101837

Mathur, V. N., Afionis, S., Paavola, J., Dougill, A. J., & Stringer, L. C. (2014). Experiences of host communities with carbon market projects: Towards multi-level climate justice. *Climate Policy*, *14*(1), 42–62. https://doi.org/10.1080/14693062.2013.861728

Megura, M., & Gunderson, R. (2022). Better poison is the cure? Critically examining fossil fuel companies, climate change framing, and corporate sustainability reports. *Energy Research & Social Science*, *85*, 102388. https://doi.org/10.1016/j.erss.2021.102388

Meta. (2021). *2020 sustainability report*. https://sustainability.fb.com/wp-content/uploads/2021/06/2020_FB_Sustainability-Report

Microsoft. (2021). *2020 environmental sustainability report*. https://cdn-dynmedia-1.microsoft.com/is/content/microsoftcorp/microsoft/

Microsoft. (2024). *2024 environmental sustainability report*. https://cdn-dynmedia-1.microsoft.com/is/content/microsoftcorp/microsoft/

Milman, O. (2021, October 1). Apple and Disney among companies backing groups against US climate bill. *The Guardian*. https://www.theguardian.com/us-news/2021/oct/01/apple-amazon-microsoft-disney-lobby-groups-climate-bill-analysis

Monsees, L., Liebetrau, T., Austin, J. L., Leander, A., & Srivastava, S. (2023). Transversal politics of Big Tech. *International Political Sociology*, *17*(1), 1–23. https://doi.org/10.1093/ips/olac020

Mwema, E., & Birhane, A. (2024). Undersea cables in Africa: The new frontiers of digital colonialism. *First Monday*, *29*(4). https://doi.org/10.5210/fm.v29i4.13637

Narain, S. (2019). Equity: The final frontier for an effective climate change agreement. In K.-K. Bhavnani, J. Foran, P. A. Kurian, & D. Munshi (Eds.), *Climate futures: Reimagining global climate justice* (pp. xxiv–xxix). Bloomsbury Publishing.

Newell, P., Srivastava, S., Naess, L. O., Torres Contreras, G. A., & Price, R. (2021). Toward transformative climate justice: An emerging research agenda. *Wiley Interdisciplinary Reviews: Climate Change*, *12*(6), e733. https://doi.org/10.1002/wcc.733

Nothias, T. (2020). Access granted: Facebook's free basics in Africa. *Media, Culture & Society*, *42*(3), 329–348. https://doi.org/10.1177/0163443719890530

O'Brien, I. (2024, September 15). Data center emissions probably 662% higher than big tech claims. Can it keep up the ruse? *The Guardian*. https://www.theguardian.com/technology/2024/sep/15/data-center-gas-emissions-tech

Oyedemi, T. D. (2020). Digital coloniality and "next billion users": The political economy of Google Station in Nigeria. *Information, Communication & Society*, *24*(3), 329–343. https://doi.org/10.1080/1369118X.2020.1804982

Pande, R. (2024). Offsets, carbon markets, and climate and economic justice. *Science*, *385*(6714), eads1902. https://doi.org/10.1126/science.ads1902

Pantti, M., Laaksonen, S.-M., & Karell, E. (2025). Platform power and climate responsibility: A longitudinal view on digital platform companies' environmental discourse. *Platforms & Society*, *2*. https://doi.org/10.1177/29768624251380428

Poell, T., Nieborg, D., van Dijck, J., Caplan, R., Helmond, A., van der Vlist, F., Chen, J., & Plantin, J.-C. (2023). *Locating and theorizing platform power*. AoIR Selected Papers of Internet Research. https://doi.org/10.5210/spir.v2022i0.12965

Pollach, I. (2018). Issue cycles in corporate sustainability reporting: A longitudinal study. *Environmental Communication*, *12*(2), 247–260. https://doi.org/10.1080/17524032.2016.1205645

Russell, A. (2023). *The mediated climate: How journalists, big tech, and activists are vying for our future*. Columbia University Press.

Sardo, M. C. (2023). Responsibility for climate justice: Political not moral. *European Journal of Political Theory*, *22*(1), 26–50. https://doi.org/10.1177/1474885120955148

Schlichting, I. (2013). Strategic framing of climate change by industry actors: A meta-analysis. *Environmental Communication*, *7*(4), 493–511. https://doi.org/10.1080/17524032.2013.812974

Schlosberg, D. (2012). Climate justice and capabilities: A framework for adaptation policy. *Ethics and International Affairs*, *26*(4), 445–461. https://doi.org/10.1017/S0892679412000615

Schlosberg, D., & Collins, L. B. (2014). From environmental to climate justice: Climate change and the discourse of environmental justice. *Wiley Interdisciplinary Reviews. Climate Change*, *5*(3), 359–374. https://doi.org/10.1002/wcc.275

Schneider, L. (2024). *COP29 Exacerbates Climate Injustice*. https://www.boell.de/en/2024/11/29/cop29-exacerbates-climate-injustice

Shawoo, Z., & McDermott, C. L. (2020). Justice through polycentricity? A critical examination of climate justice framings in Pakistani climate policymaking. *Climate Policy*, *20*(2), 199–216. https://doi.org/10.1080/14693062.2019.1707640

Shue, H. (1999). Global environment and international inequality. *International affairs*, *75*(3), 531–545. https://doi.org/10.1111/1468-2346.00092

Shue, H. (2014). *Climate justice: Vulnerability and protection*. Oxford University Press.

Shue, H. (2017). Responsible for what? Carbon producer CO_2 contributions and the energy transition. *Climatic Change*, *144*(4), 591–596. https://doi.org/10.1007/s10584-017-2042-9

Siltaoja, M. E., & Onkila, T. J. (2013). Business in society or business and society: The construction of business–society relations in responsibility reports from a critical discursive perspective. *Business Ethics, The Environment and Responsibility*, *22*(4), 357–373. https://doi.org/10.1111/beer.12028

Singh, P., & Bajwa, A. (2023). Who is responsible for the climate crisis? A perspective from the Global South. *International Critical Thought*, *13*(3), 429–444. https://doi.org/10.1080/21598282.2023.2256474

Sultana, F. (2022). The unbearable heaviness of climate coloniality. *Political Geography*, *99*, 102638. https://doi.org/10.1016/j.polgeo.2022.102638

Symbiosis Coalition. (2024). *Press release*. https://www.symbiosiscoalition.org/perspectives/launch-press-release

Umbers, L., & Moss, J. (2020). *Climate justice beyond the state*. Routledge.

UN Trade and Development. (2024). *Classifications*. https://unctadstat.unctad.org/EN/Classifications.html

UNFCCC. (1992). *The United Nations framework convention on climate change*. https://unfccc.int/resource/docs/convkp/conveng.pdf

van de Poel, I. (2015). The problem of many hands. In I. van de Poel, L. M. M. Royakkers, & S. D. Zwart (Eds.), *Moral responsibility and the problem of many hands* (pp. 50–92). Routledge.

van Dijck, J., Nieborg, D., & Poell, T. (2019). Reframing platform power. *Internet Policy Review*, *8*(2), 1–18. https://doi.org/10.14763/2019.2.1414

Young, I. M. (2006). Responsibility and global justice: A social connection model. *Social Philosophy and Policy*, *23*(1), 102–130. https://doi.org/10.1017/S0265052506060043

Young, I. M. (2011). *Responsibility for justice*. Oxford University Press.

Zuckerberg, M. (2017, May 5). *Building global community*. Facebook. https://www.facebook.com/notes/mark-zuckerberg/building-global-community/10154544292806634/

Section 4

Closing Words

Afterword: Big Tech in a World-in-Crisis

Simon Cottle

Cardiff University, UK

In these final reflections, I build on the concerns and acute empirical analyses of the contributors to *Platforms and the planet* and offer a few wide-angled thoughts on digital platforms and environmental responsibility when seen and situated within today's 'world-in-crisis' (Cottle, 2023, 2025). A context that is today evidentially both planetary in scale and existential in nature, and in which Big Tech, as we shall hear, is more deeply implicated than many may suppose – materially, economically, culturally, politically.

Big Tech information and communication corporations and platforms such as Alphabet (Google), Amazon, Apple, Meta (Facebook, Instagram, WhatsApp) and Microsoft (LinkedIn, Skype), the so-called big five, as well as Nvidia, Twitter/X and others are the digital heirs to 19th and 20th century legacy media, including press barons and media Moghul empires. Their operations straddle the globe, and they have created some of the richest and most powerful individuals on the planet. As communication infrastructures, cultural forms and daily practices, they have become deeply sutured into public life and private lives around the world. They variously shape and condition corporate and political activity, professional and cultural practices, individual identities and even our innermost psychological drives and desires. And so too are they deeply implicated in today's world-in-crisis.

We inhabit a world of systemic, deepening and interpenetrating crises where not only human existence but also the web-of-life itself has come under increasing assault and is becoming dangerously degraded. This is so much more than climate change, which itself needs to be situated, explained and responded to in relation to a world that is now overshooting its finite resources and in which the age of fossil-fuelled extractivism and rampant ecological exploitation and devastation is visibly failing and must soon come to an end. The complex of entangled crises that now pose existential threats include climate change but also bio-diversity loss and ecosystem collapse, the sixth mass extinction, zoonotic diseases and global pandemics, food, water and energy insecurity, top-soil degradation, mass migrations, world population growth and weapons of mass destruction, all exacerbated by and often experienced through failing supply

Platforms and the Planet, 203–214

doi:10.1108/978-1-83662-172-020261013

chains, rising costs of living, increasing humanitarian disasters, political polarisation, the rise of authoritarianism and new propensities for conflicts and wars. The amassed empirical evidence and expert analyses on all the above are now considerable and compelling; we are indeed living in a world-in-crisis – a world characterised by globally endemic, entangled and mutually exacerbating crises, many of them planetary in scale and existential in potential consequence (Cottle, 2025; see also Albert, 2024; Bendell, 2023; Miller & Heinberg, 2023; Office for the Coordination of Humanitarian Affairs, 2024; United Nations University, Environment and Human Security, 2023; United Nations Environment Program, 2024; World Wide Fund for Nature, 2024).

The recent proliferation of terms for this world (dis)order speaks to the awakening need to better perceive and respond to the converging trajectories of destabilisation and potential, possible and probable collapse. Whether conceived, for example, in terms of a 'world risk society' (Beck, 2000), 'planetary emergency' (Club of Rome, 2019; Thunberg, 2023), 'civilisational collapse' (Bendell, 2023; Read & Alexander, 2019), 'global polycrisis' (Albert, 2024; Miller & Heinberg, 2023; World Economic Forum, 2023), 'metacrisis' (Hagens, 2024) and 'world-in-crisis' (Cottle, 2023; Lake, 2024), they all, among others, signal the historically unprecedented and catastrophic nature of our anthropogenically induced planetary demise. This includes ecology and ecosystems and not just the 'environment' as ordinarily understood. And it is also planetary and not just global or world in scope because the dynamics of destabilisation and degeneration in the era of the Anthropocene (Crutzen & Steffen, 2003; Steffen et al., 2015), or Capitalocene (Moore, 2015), are now impacting the very life support systems of planet Earth, including its encompassing biosphere (Foster, 2022; Lenton et al., 2023; World Wide Fund for Nature, 2024).

This is not to suggest that we do not also exist in a cultural pluriverse and multi-polar political world or that the gross inequalities both between and within the Global North and Global South do not profoundly impact life chances and indeed the chance of life itself (Beck, 2006; Chakrabarty, 2021; Escobar, 2020). World inequality and differing cultural outlooks and cosmologies condition how extreme precarity becomes experienced, understood and responded to. This is true, but our world has now become universally ensnared by human history's most globally rapacious, economically extractive and ecologically devastating system of production and consumption yet set loose on the planet. A seemingly insatiable capitalist system that is underpinned by a normative worldview wedded to incessant growth, materialist aspirations of progress and the enlightenment belief in human exceptionalism and speciesism that hubristically positions human beings and their enterprise as if they were outside of nature and in control of the web-of-life.

Big Tech and Their Platforms: What is the Problem?

So, what does all this have to do with Big Tech? I could simply observe that I am writing this on my Apple Mac, making use of Google Chrome and the Google search facility to access pertinent websites and retrieve the latest reports documenting today's world-in-crisis. On my iPhone and via Google's search engine, I have recently accessed Amazon UK's book department and bought, for example, Sunil Amrith's *The burning earth* (2024), Robin Wall Kimmerer's *The serviceberry* (2024) and Richard Seymour's *Disaster nationalism: The downfall of liberal civilization* (2024), all available in Kindle formats.

My attention has also sporadically been drawn to various political and ecological posts on Facebook, and I have received countless emails via Microsoft Outlook from a wide array of online protest movements organising various campaigns. In the wake of President Trump's second election victory and his divisive policies and disastrous 'Drill, baby, drill' climate intentions, these included, for example, collective emails from: Global Citizen, who emailed today an Action Alert: 'To take action to defeat poverty, defend the planet & demand equity!', the Global Compassion Coalition who asked, 'Want to know what's in store for us this year?' and from the radical collective Eko who invited me to 'sign a petition banning Elon Musk from pouring money into and buying the next UK election'.

And just this very minute, an email from Resilience.org, melodically pinged into my Microsoft Outlook inbox taking me to their online 'Weekly society digest'. This includes articles with the opening lines 'To confront the oligarchy, we need to build power at the community level', and 'How archaeologists can solve the earth's "wicked problems"'. Resilience.org, produced by the Post-Carbon Institute, further invited me to share these and other articles on Facebook, tweet via X and through my email. I also received a collective email from Nate Hagens, from The Great Simplification, alerting me to his latest podcast, 'Power vs life: Toward wide boundary sovereignty', found on his website that hosts numerous filmed interviews, accessed via YouTube, with many of the world's leading Earth scientists, philosophers and cutting-edge thinkers all contemplating today's metacrisis.

In such multiple, technologically connecting and interacting ways I am personally, it seems, daily immersed in and communicatively dependent on Big Tech and its platforms – though I suspect younger generations will make even more use of an expanded array of communication devices and platforms than this. So, what is the problem, some might reasonably ask. Are not these Big Tech infrastructures and platforms thereby communicating today's world cacophony of ecological and political voices and viewpoints? Are they not facilitating connectivity, the cultural commons and helping to build compassion and organised concern around the planet? Unfortunately, I do not think they are and for three fundamental reasons, each addressed below.

System Dominance and Digital Dissimulation: Amusing Ourselves to Death 2.0

Big Tech corporations are some of the biggest corporations on the planet, capitalising from processes of technological convergence, globalisation, deregulation and financialisation. Whether conceived and theorised in terms of 'technofeudalism' (Varoufakis, 2024) or 'platform capitalism' (Gane, 2024; Gilbert, 2024; Snick, 2017), they remain no less wedded to constant growth and the stimulation of consumer demands than earlier transnational corporations. Though their near monopoly power, multiple platforms and mastery of algorithmically targeted campaigns and messaging affords them a huge market (and political) advantage. They have become key structural pillars supporting today's global capitalist marketplace, and they are economically determined by this same marketplace to compete, innovate and grow.

And so too are they thereby obliged to extract and exploit precious Earth minerals and materials and use energy in the design, manufacture, promotion, selling and transportation of their devices, products and platforms and in the maintenance of their ubiquitous technological infrastructures. The latter now includes multiple data centres and energy-intensive cloud storage systems that are expected to expand exponentially in the years ahead to keep up with new communications developments such as generative artificial intelligence (AI). As many of the contributors to *Platforms and the planet* have incisively analysed, and notwithstanding Big Tech's greenwashing claims and declared commitments to sustainability, these corporations operate within today's global, heavily financialised and deregulated world markets, building products with built-in-obsolescence and using scarce minerals and metals (e.g. indium, tantalum, gallium and rare earth elements lithium, cobalt, germanium and tellurium) and end-of-life polluting materials (e.g. lead, mercury and arsenic) that can leach into the subsoil and water courses.

Big Tech, as the generic term implies and corporate behaviours demonstrate, are all about market dominance both in terms of their own corporate positioning and ambitions but also in terms of their underpinning infrastructural role in the global capitalist market system. This infrastructural importance of Big Tech is not lost on politicians and world leaders who are now, for example, investing billions into the AI revolution, seeking competitive advantage while harnessing its undoubted potential to national and political goals. This includes the current British Labour Government's AI Opportunities Plan (Department for Science, Innovation & Technology, 2025), designed to boost growth in public services and the economy more widely as part of its national economic mission and in support of the government's oft-stated mantra of 'Growth, Growth, Growth'. President Trump, prompted by the release of the Chinese DeepSeek AI system and its impact on US Big Tech stock values, is also in no doubt of Big Tech's importance for the continuing dominance of US society and economic system: 'Under the Trump administration', he boasts, 'we're going to unleash our tech companies and we're going to dominate the future like never before' (Dodds, 2025).

The power of Big Tech is not only economic, in the traditional political economy sense of corporate competitive behaviour oriented to profit maximisation and share-holder dividends and won principally through growth, synergies of convergence and monopolisation; the latter driving down costs and maximising product and service sales. It is also cultural. Big Tech infrastructures and platforms are purveyors of meanings, values, narratives and even dreams as well as facilitators of connectivity and sometimes, solidarities and imagined communities. My personal communicative dependency and immersion in Big Tech and an incoming array of alternative ecological and political viewpoints is not, of course, representative of everyone's, or indeed anyone else's communication consumption patterns. Personal choices and past communication behaviours as well as tailored algorithms and anterior corporate push are all at play in everyday Big Tech engagement. Here issues of cultural capital and unequal access around the globe can effectively splinter and fragment the online 'cultural commons' and disempowers those who cannot meaningfully participate, deliberate with others and forge a sense of solidarity or imagined community.

My personal immersion in Big Tech and platforms also tells me, however, that it is impossible to avoid the daily avalanche of unsolicited corporate advertising and promotions including trailers and promos for streaming platforms and mainstream entertainment products, political click-bait requests for support and unsolicited personal communications veering towards the fraudulent and sometimes criminal. Toxic pollution, it seems, is not confined to our surrounding environments of land, water and air but also colonises and corrodes our screen-based mental habitus, sometimes with the explicit politics of hate and division, and more habitually in and through the generalised banalities and normative rehearsals of so much that is 'entertaining' in contemporary culture. Some time ago, in a different media universe, when television was still king, Neil Postman dissected the problems of the medium, its epistemology in comparison to typographic communications, its entertainment forms and trivialising appeals and how American society allowed this to become its cultural fulcrum – and at what social and political costs (Postman, 1987). Culture-death and the undermining of public discourse were at the core of Postman's telling indictment on American television that he termed 'Amusing ourselves to death'.

A thoroughgoing analysis of Big Tech and its different platforms is beyond the scope of these few pages, and I am mindful of not simply asserting a monolithic view of Big Tech cultural dominance, though there is no doubting its economic and infrastructural predominance at the heart of today's neo-liberal world marketplace. Alternative and oppositional voices and viewpoints contesting within wider civil and global society, and struggling for ecological and societal change, as intimated, can and do find a foothold in and through the infrastructures and platforms of Big Tech companies. And this should not be under-estimated and needs to be deliberatively deepened and expanded in the future. But this remains, to borrow an Althusserian phrase, 'structured in dominance', with alternative and oppositional voices effectively positioned at the periphery and struggling to secure more widespread representation and connection with wider publics who are themselves predominantly positioned and addressed as consumers by so much of Big Tech's carrying capacity, connectivity and content.

Today's Big Tech platforms offer a wealth of distractions and seductions to communicatively hyper-individualised consumers. This includes the constant succession of, and scrolling for, visually arresting dopamine-hit images and fragmentary human-interest stories, as well as the narcissistic interventions of influencers, the ill-informed and opinionated, and the seducing algorithmic push of pornography and dating-sites. Corporate on-line advertising, on-demand streaming of mass entertainment products and professional and personal networks are all in the mix, as are major news organisations, political parties and their online campaigns, and the charity appeals of civil society organisations. All this can present sometimes as a radically fragmented and cacophonous world, but a world that is individually entered, habitually accommodated and algorithmically tailored and reproduced. A world with interjecting issues, contending identities and conflicts certainly, but a world without real-world contradictions such as those between economic growth and ecological exhaustion, solidarities of compassion and the self-centred pursuit of consumption, and humankind's presumed dominion over nature rather than custodianship and symbiosis inside the web-of-life.

And therein lies the core problem when set against the looming backdrop of our world-in-crisis. In a world of endemic, overlapping and deepening existential crises, Big Tech's communicative world generally reads as one of habitual integration in the systems world of 'business as usual', and in the lifeworld as one of privatised distraction and dissimulation. This is a far cry from the ideal of the technologically instantiated and collectively enacted cultural commons.

Big Tech, Political Authoritarianism and the Demagogue's Playground

Big Tech enters today's world-in-crisis not only through its economic powerhouse activities, its core infrastructural role in global financialised capitalism and its platforms' normative endorsement of an unsustainable consumer culture. It also increasingly enters the political turmoil unleashed by a world-in-crisis and the generation of growing ecological destabilisation and economic duress. Social media, the Internet and other platforms have all been put to work in the recent rise of far-right authoritarian politics in Europe, America and beyond, spreading lies and fomenting hate, and algorithmically targeting demographic subgroups with tailored psychologistic messages designed to recruit support and win elections.

This communicative field has effectively become the demagogue's playground. Misinformation and disinformation, propaganda and lies have long been part of the playbooks of authoritarian leaders and far-right extremists. This can more easily flourish in times of political polarisation, crisis and confusion (Arendt, 2022/1971). We also know that some groups of people – the disempowered and disenfranchised, the alienated and angry, the poor and uneducated – have less resilience and are more susceptible to such messaging. Authoritarian leaders have long plied conspiracy theories when blaming outgroups for society's ills, labelling them as enemies while liberal democratic norms and safeguards are dismantled. There is an unfortunate elective affinity between Big Tech's social media and platforms and its algorithmically generated filter bubbles and self-selected echo-chambers. Both are seemingly anaemic to deliberative democratic politics while reinforcing preferred views and beliefs and encouraging processes of social polarisation (Seymour, 2024). Their platforms recently circulated conspiracy claims about government manufacture of climate change events, for example, and literally granting political platforms to extremist, anti-democratic views.

Would-be demagogues also benefit from the new digital communications environment through the so-called 'liar's dividend' (Chesney & Citon, 2019). Here the utterer of a politically hateful statement in the media, for example, who is then called out for it by other media, may simply disown their original statement, saying it was not them, it was a media distortion or perhaps a malicious deep fake. In this way, the original hateful message is repeated and gains increased publicity and even political traction in the ensuring media furore or scandal, while the perpetrator defends their position of innocence and evades generalised condemnation. The 'liar's dividend' is part of the wider breakdown in trust in political and media communications that fuels the demagogue's communication playground, enabling bad political actors to stoke confusion, deny blatant wrongdoing and divide political opponents. It is part of today's digitally emerging and democratically injurious epistemological crisis that can benefit authoritarian leaders seeking to dismantle democratic processes and evade their checks and balances. A crisis in which people lose faith in all truth claims or find it hard to adjudicate between those that are false and those that are true. In times of epistemological crisis, as Hannah Arendt astutely observed in the last century, politics can lurch surprisingly quickly from liberal democratic norms towards fascism and totalitarianism.

> The result of a consistent and total substitution of lies for factual truth is not that the lie will now be accepted as truth and truth be defamed as a lie, but that the sense by which we take our bearings in the real world – and the category of truth versus falsehood is among the mental means to this end – is being destroyed. (Arendt, 2022/1971, p. 47)

Early on in his first presidency, Donald Trump deliberately labelled the mainstream news media as 'fake news' and 'fake media' and did so on average more than once every day, and more so when responding to the media's fact-checking of his tirade of lies, constant defamations of character and denials of countless wrongdoings (Stelter, 2018). In a climate of rising political extremism, including Trump's own interventions, social media platforms have come under increasing pressure to moderate content and bolster their fact-checking provision and censor politically hateful and potentially violent posts. With President Trump's return to the White House in 2025, this democratic safeguard came under assault.

Mark Zuckerberg, CEO Meta (incorporating Facebook and Instagram), suspended the formally independent fact checking programme. Previously Facebook and Instagram had suspended Trump's own accounts for two years, following his praise for people involved in violence at the Capitol Hill on 6 January 2021. Following his re-election, Zuckerberg donated $1 million to Trump's inaugural fund and appointed his prominent supporter Dana White,

CEO of Ultimate Fighting Championship, to Meta's Board of Directors. Elon Musk, CEO of Twitter/X, for his part, has repeatedly used his ownership of the platform to propagate misinformation and disinformation, including making incendiary claims fuelling anti-asylum seeker riots in the United Kingdom, uncritically showcasing Donald Trump's views during his second election period, and hosting a 74-minute live chat with Alice Weidel, spokesperson for the Alternative fur Deutschland (AfD), Germany's far-right extremist party. He also directed for a time President Trump's new Department of Government Efficiency (DOGE) terminating countless government jobs and social and environmental programs.

The fanciful idea that Big Tech is somehow different in its market behaviour and political outlooks because of its entrepreneurial leaders or roots in Silicon Valley and earlier Californian hippy-culture is blown away by the extent to which CEOs like Zukerberg have personally capitulated to, and CEOs like Musk have enthusiastically supported, President Trump's polarising political and media agenda. It is telling perhaps that Meta's Zuckerberg, Amazon's Jeff Bezos, Google's Sundar Pichai, Apple's Tim Cook and TikTok's Zhou Zi Chew all attended President Trump's inauguration, along with senior oil and gas industry figures and the traditional right-wing press baron, Rupert Murdoch. In today's world-in-crisis, Trump's nationalistic 'Make America great again', ecologically catastrophic 'Drill, baby drill' politics and de-regulatory environmental and food policies, among others, have clearly not only been aided and abetted by Big Tech companies but have been enthusiastically and irresponsibly endorsed.

Going Deeper: Communicative Ecology and the Web-of-Life

To situate and better understand the responsibility and failings of Big Tech companies and platforms in today's world-in-crisis, I have argued we must attend to not only their environmental and carbon footprints and greenwashing practices but also to their turbocharging infrastructural role in the global economic system premised on incessant unsustainable growth. And so too must we examine the extent to which they communicatively register and culturally respond to today's world of ecological degradation and social injustice. To grasp the enormity and interrelatedness of the multiple global crises that now deepen and interpenetrate in an ecologically overshooting and politically destabilising world, this requires a holistic and not a disaggregated single-issue approach. When we begin to see the bigger picture of an unsustainable world seemingly locked into unstoppable economic growth (and underpinned by historically unprecedented financialised debt), and how different crises – climate change, biodiversity loss, water scarcity, food insecurity, deforestation, zoonotic diseases, conflicts, mass migrations and others – are negative symptoms of human society's insatiable economic system and its underpinning drivers, so we also begin to glimpse in reverse image positive ecologically sustainable and socially just alternatives.

Today's world-in-crisis can be seen as an interregnum between the age of fossil-fuelled industrialism and ecological extraction and exploitation, which is now visibly dying and a new age, still inchoate and barely glimpsed, but necessary if human life and the biosphere is to have a chance of survival into the future. This is sometimes given the name of 'ecological civilisation' (Korton, 2021; Lent, 2021) or 'the Symbiocene' (Albrecht, 2019) and is reflected around the world in a broad constellation of different views and voices, traditions and practices. Its many disparate currents include, for example, the spiritualities of Taoism and Buddhism (Lent, 2021; Loy, 2019; Matthews, 2023), the sensibilities of nature found in Romanticism (Sayre & Lowy, 2021) and Deep Ecology (Schumacher, 2011/1971; Naess, 2021), as well as the new 'ecological thinking' in the new post-Enlightenment philosophy of science (Capra & Luisi, 2014; McGilchrist, 2021; Theise, 2023) and complex systems (Meadows, 2008; Wahl, 2016). And encompassed here are also the cosmologies and wisdom of indigenous peoples (Kimmerer, 2025; Lake, 2024), the alternative and regenerative practices of New Age lifestyles and new social movements struggling against social injustice and planetary ecological degradation. For the most part, they all centre on a less materialistic, more regenerative, more local, more reciprocal and ecologically symbiotic and sustainable way of life and are respectful of our own part in nature and the web-of-life.

At their heart, these diverse traditions, beliefs and practices share an 'ecological' understanding of the web-of-life: its wholeness and relationality, not boundedness or atomisation into parts; its complex systems and nested and fractal structures and the capacity of organisms to adapt, evolve and exist in symbiosis. Interconnectedness and interdependency are emphasised over possessive individualism, competition and control, and this includes one's own body and self-consciousness when viewed as expressive and momentary in nature and in flux in the wider cycles of life. This 'ecological' understanding can assume a religious or spiritual form or a secular and scientific one and offers a way of making sense of the unmaking of the contemporary world, the biosphere and prospects of future life on planet Earth.

This cultural flourishing around the globe, sometimes called 'the Awakening' or 'the Great Turning' (Macy, 2021), in response to the discernible 'Great Unravelling', is characteristically motivated to find a more harmonious, less exploitative way of living with others and in custodial relationship to nature and the biosphere. It does so in the teeth of a globalised world seemingly set on self-destruction and which does not have the historical and future imagination or, currently, the political and moral will to seriously engage with the necessary project of transformational change. It encompasses and explores imagined futures and new cultural designs for living, from circular economies, localism and degrowth to bioregions, regeneration and political subsidiarity and so much else besides. It is broadly humanistic, compassionate and critically engaged.

This emergent cultural constellation warrants a prominent platform and ubiquitous presence in and through Big Tech. It deserves to be heard, listened to, deliberated and democratically acted upon. It speaks to today's world-in-crisis, its underpinning materialist and ideational drivers, and it proffers ecologically sustainable and socially just ways of living in a shared, survivable world. In this way, it offers biophilic resources of hope in a visibly dying world. This cultural flourishing of ideas and practices, of critique and imagined futures, as well as the politics of active hope has yet, however, to be fully recognised or find centre stage on Big Tech's fractured 'cultural commons' and dissimulating world of Amusing ourselves to death 2.0. or within its deliberatively allergic demagogue's playground of social media, partisan platforms and self-referential echo-chambers.

Concluding Words

Gaia as well as corporate capitalism, it seems, desperately needs the devil's tools. The global reach, planetary consequence and communicative cacophony of Big Tech and its digital platforms are predominantly weighted and enacted in ways that serve 'business as usual'. A worldview or social imaginary that is premised on economic growth, materialist ideas of progress and human exclusivism and speciesism and that has brought the planet to its current demise, and, possibly, end-times. And yet, something akin to Big Tech is surely needed in a cultural pluriverse and multipolar political world that desperately needs to confer and act in concert to alleviate or mitigate some of the worst ecological and inhumane scenarios now fast coming into view. The cultural flourishing we see around the world today, in part set lose by the accelerating crush of climate disasters, ecological degradations and social injustices, needs more, not less, Big Tech and technologically innovative platforms.

The emergent and converging constellation of ecological ways of thinking and imagined future ways of being must take root in the present before they can become beacons illuminating pathways to the future. Big Tech and its creative capacity for technologically innovative platforms and global-local networks of connectivity already provide a powerful foothold for new social movements and those struggling to affect ecological and social change. But as we have heard, so too is this 'contained' by Big Tech corporate imperatives, the digital distractions and the deliberative vacuum and epistemological uncertainties found in the demagogue's playground of social and partisan media. Challenges abound in the world of oft-supposed communicative abundance.

The radical flank of environmental consciousness, as well as the mainstream waking up to a world-in-crisis, need to assemble and deliberate on the cultural commons of Big Tech connectivity and platform representation, and when and if they do, this could yet become a force multiplier for transformation and change in alignment with the web-of-life. Though urgently required, the movement for change will first have to negotiate and overcome some of the obstacles and challenges of Big Tech that is currently positioned deep inside and exacerbating today's world-in-crisis.

References

Albert, M. J. (2024). *Navigating the polycrisis*. MIT Press.

Albrecht, G. (2019). *Earth emotions.* Cornell University Press.

Amrith, S. (2024). *The burning earth: An environmental history of the last 500 years.* Allen Lane.

Arendt, H. (2022/1971). *On lying and politics.* Library of America.

Beck, U. (2000). *World risk society.* Polity.

Beck, U. (2006). *Cosmopolitan vision.* Polity.

Bendell, J. (2023). *Breaking together.* Good Works.

Capra, F., & Luisi, P. L. (2014). *The systems view of life.* Cambridge University Press.

Chakrabarty, D. (2021). *The climate of history in a planetary age.* University of Chicago Press.

Chesney, B., & Citon, D. (2019). Deep fakes: A looming challenge for privacy, democracy and national security. *California Law Review, 107*, 1753–1820.

Club of Rome. (2019). *Planetary emergency plan.* https://www.clubofrome.org/publication/the-planetary-emergency-plan/

Cottle, S. (2023). Reporting civilisational collapse: Research notes from a world-in-crisis. *Global Media and Communication, 19*(2), 269–288.

Cottle, S. (Ed.). (2025). *Communicating a world-in-crisis.* Peter Lang.

Crutzen, P., & Steffen, W. (2003). How long have we been in the Anthropocene era? *Climatic Change, 61*(3), 251–257.

Department for Science, Innovation & Technology. (2025). *AI opportunities plan.* (Crown publication number CP1241). https://www.gov.uk/government/publications/ai-opportunities-action-plan/ai-opportunities-action-plan

Dodds, I. (2025, January 28). Trump warns new Chinese DeepSeek AI is a "wake up call" for US tech companies. *Independent.* https://www.independent.co.uk/news/world/americas/us-politics/donald-trump-deepseek-ai-china-b2687418.html

Escobar, A. (2020). *Pluriversal politics.* Duke University Press.

Foster, J. (2022). *Capitalism in the Anthropocene.* Monthly Review Press.

Gane, N. (2024). Capitalism is capitalism, not technofeudalism. *Journal of Classical Sociology, 25*(4), 353–368.

Gilbert, M. (2024). Techno-feudalism or platform capitalism? Conceptualising the digital society. *European Journal of Social Theory, 27*(4), 561–578.

Hagens, N. (2024, May 8). *Confronting the metacrisis: Reality roundtable 9 [video].* YouTube. https://www.youtube.com/watch?v=SnpsxGQ0aZY

Kimmerer, R. W. (2025). *The serviceberry: An economy of gifts and abundance.* Allen Lane.

Korton, D. (2021). *Ecological civilization: From emergency to emergence* [White paper]. https://davidkorten.org/wp-content/uploads/2021/11/Korten-EcoCiv-11032021.pdf

Lake, O. (2024). *The story is in our bones: How worldviews and climate justice can remake a world in crisis.* New Society Publisher.

Lent, J. (2021). *The web of meaning.* Profile Books.

Lenton, T., McKay, D., Loriani, S., Abrams, J., Lade, S., Donges, J., Milkoreit, M., Powell, T., Smith, S., Zimm, C., Buxton, J., Bailey, E., Laybourn, L., Ghadiali, A., & Dyke, J. (Eds.) (2023). *The global tipping points report 2023.* University of Exeter. https://global-tipping-points.org/

Loy, D. R. (2019). *Ecodharma.* Wisdom Publications.

Macy, J. (2021). *World as lover, world as self.* Parallax Press.

Matthews, F. (2023). *The Dao of civilisation.* Anthem Press.

McGilchrist, I. (2021). *The matter with things.* Perspectiva Press.

Meadows, D. (2008). *Thinking in systems*. Chelsea Green Publishing.
Miller, A., & Heinberg, R. (2023). *Welcome to the Great Unraveling: Navigating the polycrisis of environmental and social breakdown*. Post Carbon Institute.
Moore, J. (2015). *Capitalism in the web of life*. Verso Books.
Naess, A. (2021). *There is no point of return*. Penguin.
Office for the Coordination of Humanitarian Affairs. (2024). *Global humanitarian overview 2024*. https://www.unocha.org/publications/report/world/global-humanitarian-overview-2024-enarfrsp
Postman, N. (1987). *Amusing ourselves to death*. Methuen.
Read, R., & Alexander, S. (2019). *This civilization is finished*. Simplicity Institute.
Sayre, R., & Lowy, M. (2021). *Romantic anti-capitalism and nature*. Routledge.
Schumacher, E. F. (2011/1971). *Small is beautiful. A study of economics as if people mattered*. Vintage books.
Seymour, R. (2024). *Disaster nationalism: The downfall of liberal civilization*. Verso.
Snick, N. (2017). *Platform capitalism*. Polity Press.
Steffen, W., Broadgate, W., Deutsch, L., Gaffney, O., & Ludwig, C. (2015). The trajectory of the Anthropocene: The Great Acceleration. *The Anthropocene Review*, *2*(1), 81–98.
Stelter, B. (2018, January 17). *Trump averages a "fake" insult every day. Really. We counted*. CNN Business. https://money.cnn.com/2018/01/17/media/president-trump-fake-news-count/index.html
Theise, N. (2023). *Notes on complexity*. Spiegel and Grau.
Thunberg, G. (Ed.). (2023). *The climate book*. Penguin Random House.
United Nations Environment Programme. (2024). *Emissions gap report*. https://www.unep.org/resources/emissions-gap-report-2024
United Nations University, Environment and Human Security. (2023). *Interconnected disasters risk report 2023*. https://interconnectedrisks.org
Varoufakis, Y. (2024). *Technofeudalism*. Vintage.
Wahl, D. C. (2016). *Designing regenerative cultures*. Triarchy Press.
World Economic Forum. (2023). *We're on the brink of a polycrisis, how worried should we be? Global risks report 2023*. https://www3.weforum.org/docs/WEF_Global_Risks_Report_2023.pdf
World Wide Fund for Nature. (2024). *World Wide Fund for nature living planet report 2024*. https://livingplanet.panda.org/en-GB/

Index

Printed and bound by CPI Group (UK) Ltd, Croydon, CR0 4YY

28/05/2026

14888784-0002